# The East Bay Hills Fire
# Oakland-Berkeley, California

Investigated by: J. Gordon Routley

This is Report 060 of the Major Fires Investigation Project conducted by TriData Corporation under contract EMW-90-C-3338 to the United States Fire Administration, Federal Emergency Management Agency.

Department of Homeland Security
United States Fire Administration
National Fire Data Center

# U.S. Fire Administration Fire Investigations Program

The U.S. Fire Administration develops reports on selected major fires throughout the country. The fires usually involve multiple deaths or a large loss of property. But the primary criterion for deciding to do a report is whether it will result in significant "lessons learned." In some cases these lessons bring to light new knowledge about fire--the effect of building construction or contents, human behavior in fire, etc. In other cases, the lessons are not new but are serious enough to highlight once again, with yet another fire tragedy report. In some cases, special reports are developed to discuss events, drills, or new technologies which are of interest to the fire service.

The reports are sent to fire magazines and are distributed at National and Regional fire meetings. The International Association of Fire Chiefs assists the USFA in disseminating the findings throughout the fire service. On a continuing basis the reports are available on request from the USFA; announcements of their availability are published widely in fire journals and newsletters.

This body of work provides detailed information on the nature of the fire problem for policymakers who must decide on allocations of resources between fire and other pressing problems, and within the fire service to improve codes and code enforcement, training, public fire education, building technology, and other related areas.

The Fire Administration, which has no regulatory authority, sends an experienced fire investigator into a community after a major incident only after having conferred with the local fire authorities to insure that the assistance and presence of the USFA would be supportive and would in no way interfere with any review of the incident they are themselves conducting. The intent is not to arrive during the event or even immediately after, but rather after the dust settles, so that a complete and objective review of all the important aspects of the incident can be made. Local authorities review the USFA's report while it is in draft. The USFA investigator or team is available to local authorities should they wish to request technical assistance for their own investigation.

This report and its recommendations were developed by USFA staff and by TriData Corporation, Arlington, Virginia, its staff and consultants, who are under contract to assist the USFA in carrying out the Fire Reports Program.

The USFA greatly appreciates the cooperation and information received from Fire Chief P. Lamont Ewell and many of the officers and firefighters of the Oakland Fire Department. Those who provided special assistance to USFA's investigation are listed on page one of this report. Appreciation also goes to Fire Chief Gary Cates, Berkeley Fire Department, Assistant Chief Bill Cullen, Contra Costa County Fire Protection District; the California Department of Forestry and Fire Protection; the California Governor's Office of Emergency Services and Mr. Bill Patterson, Federal Emergency Management Agency (Region IX Office). Assistance with the investigation was provided by Mr. Hugh Graham.

For additional copies of this report write to the U.S. Fire Administration, 16825 South Seton Avenue, Emmitsburg, Maryland 21727. The report is available on the USFA Web site at http://www.usfa.dhs.gov/

# U.S. Fire Administration

## Mission Statement

*As an entity of the Federal Emergency Management Agency (FEMA), the mission of the U.S. Fire Administration (USFA) is to reduce life and economic losses due to fire and related emergencies, through leadership, advocacy, coordination, and support. We serve the Nation independently, in coordination with other Federal agencies, and in partnership with fire protection and emergency service communities. With a commitment to excellence, we provide public education, training, technology, and data initiatives.*

# TABLE OF CONTENTS

OVERVIEW OF THE FIRE. . . . . . . . . . . . . . . . . . . . . . . . . . . . . . . . . . . . . . . 1

SUMMARY OF KEY ISSUES . . . . . . . . . . . . . . . . . . . . . . . . . . . . . . . . . . . . 2

LOCATION . . . . . . . . . . . . . . . . . . . . . . . . . . . . . . . . . . . . . . . . . . . . . . . . 4

BACKGROUND . . . . . . . . . . . . . . . . . . . . . . . . . . . . . . . . . . . . . . . . . . . . . 5

    Climatic Conditions . . . . . . . . . . . . . . . . . . . . . . . . . . . . . . . . . . . . . . 5

    Vegetation. . . . . . . . . . . . . . . . . . . . . . . . . . . . . . . . . . . . . . . . . . . . . . 6

    Land Development on the Hills. . . . . . . . . . . . . . . . . . . . . . . . . . . . . 7

    Burning Characteristics. . . . . . . . . . . . . . . . . . . . . . . . . . . . . . . . . . . 8

    Previous Fires In The East Bay Hills . . . . . . . . . . . . . . . . . . . . . . . . 11

    Other Wildland-Urban Interface Fires in California. . . . . . . . . . . . . 12

    Regulatory Efforts. . . . . . . . . . . . . . . . . . . . . . . . . . . . . . . . . . . . . . 13

    Fire Protection Agencies . . . . . . . . . . . . . . . . . . . . . . . . . . . . . . . . . 14

HOW THE FIRE STARTED . . . . . . . . . . . . . . . . . . . . . . . . . . . . . . . . . . . 15

    Point of Origin . . . . . . . . . . . . . . . . . . . . . . . . . . . . . . . . . . . . . . . . 15

    Saturday, October 19th . . . . . . . . . . . . . . . . . . . . . . . . . . . . . . . . . . 15

SUNDAY, OCTOBER 20TH . . . . . . . . . . . . . . . . . . . . . . . . . . . . . . . . . . . 17

    The Restart . . . . . . . . . . . . . . . . . . . . . . . . . . . . . . . . . . . . . . . . . . . 17

    Major Flare-Up . . . . . . . . . . . . . . . . . . . . . . . . . . . . . . . . . . . . . . . . 18

    Communications Problems . . . . . . . . . . . . . . . . . . . . . . . . . . . . . . . 21

    Critical Period. . . . . . . . . . . . . . . . . . . . . . . . . . . . . . . . . . . . . . . . . 22

    Mutual Aid Begins to Arrive . . . . . . . . . . . . . . . . . . . . . . . . . . . . . . 25

    Lives Saved And Lost . . . . . . . . . . . . . . . . . . . . . . . . . . . . . . . . . . . 26

    All Forces Retreating . . . . . . . . . . . . . . . . . . . . . . . . . . . . . . . . . . . 29

    Fire Jumps Freeway. . . . . . . . . . . . . . . . . . . . . . . . . . . . . . . . . . . . . 29

    Hiller Highlands . . . . . . . . . . . . . . . . . . . . . . . . . . . . . . . . . . . . . . . 29

    Berkeley Front. . . . . . . . . . . . . . . . . . . . . . . . . . . . . . . . . . . . . . . . . 30

    CDF Operations . . . . . . . . . . . . . . . . . . . . . . . . . . . . . . . . . . . . . . . 31

    Additional Strike Teams. . . . . . . . . . . . . . . . . . . . . . . . . . . . . . . . . . 33

    Claremont Hotel. . . . . . . . . . . . . . . . . . . . . . . . . . . . . . . . . . . . . . . 34

    New Outbreaks. . . . . . . . . . . . . . . . . . . . . . . . . . . . . . . . . . . . . . . . 35

    Strategy — Continuing Battle on Multiple Fronts . . . . . . . . . . . . . . . 36

*continued on next page*

# Table of Contents (continued)

Evacuations. . . . . . . . . . . . . . . . . . . . . . . . . . . . . . . . . . . . . . . . . . . . . . . . . . 36

Tactics. . . . . . . . . . . . . . . . . . . . . . . . . . . . . . . . . . . . . . . . . . . . . . . . . . . . . . 38

Assessment of the Situation. . . . . . . . . . . . . . . . . . . . . . . . . . . . . . . . . . . . . 39

Unified Command Structure. . . . . . . . . . . . . . . . . . . . . . . . . . . . . . . . . . . . 40

Additional Resources Ordered . . . . . . . . . . . . . . . . . . . . . . . . . . . . . . . . . . . 42

Wind Changes . . . . . . . . . . . . . . . . . . . . . . . . . . . . . . . . . . . . . . . . . . . . . . . 42

**MONDAY THROUGH THURSDAY** . . . . . . . . . . . . . . . . . . . . . . . . . . . . . 44

Losses . . . . . . . . . . . . . . . . . . . . . . . . . . . . . . . . . . . . . . . . . . . . . . . . . . . . . . 44

**ANALYSIS** . . . . . . . . . . . . . . . . . . . . . . . . . . . . . . . . . . . . . . . . . . . . . . . . . . 45

Fire Risk . . . . . . . . . . . . . . . . . . . . . . . . . . . . . . . . . . . . . . . . . . . . . . . . . . . . 45

Fire Origin . . . . . . . . . . . . . . . . . . . . . . . . . . . . . . . . . . . . . . . . . . . . . . . . . . 48

Fire Characteristics . . . . . . . . . . . . . . . . . . . . . . . . . . . . . . . . . . . . . . . . . . . 48

Wildland-Urban Interface Characteristics . . . . . . . . . . . . . . . . . . . . . . . . . 50

Training And Preparation . . . . . . . . . . . . . . . . . . . . . . . . . . . . . . . . . . . . . . 50

Incident Management . . . . . . . . . . . . . . . . . . . . . . . . . . . . . . . . . . . . . . . . . 51

Communications . . . . . . . . . . . . . . . . . . . . . . . . . . . . . . . . . . . . . . . . . . . . . 55

Public Information . . . . . . . . . . . . . . . . . . . . . . . . . . . . . . . . . . . . . . . . . . . 59

Mutual Aid . . . . . . . . . . . . . . . . . . . . . . . . . . . . . . . . . . . . . . . . . . . . . . . . . 59

Volunteer Response. . . . . . . . . . . . . . . . . . . . . . . . . . . . . . . . . . . . . . . . . . . 60

Aircraft Operations . . . . . . . . . . . . . . . . . . . . . . . . . . . . . . . . . . . . . . . . . . . 61

Stress . . . . . . . . . . . . . . . . . . . . . . . . . . . . . . . . . . . . . . . . . . . . . . . . . . . . . . 64

Firefighter Safety. . . . . . . . . . . . . . . . . . . . . . . . . . . . . . . . . . . . . . . . . . . . . 66

Emergency Medical Services (EMS). . . . . . . . . . . . . . . . . . . . . . . . . . . . . . 67

Evacuation . . . . . . . . . . . . . . . . . . . . . . . . . . . . . . . . . . . . . . . . . . . . . . . . . . 68

**LESSONS LEARNED**. . . . . . . . . . . . . . . . . . . . . . . . . . . . . . . . . . . . . . . . . . 69

**APPENDIX A:  Reference Publications** . . . . . . . . . . . . . . . . . . . . . . . . . . . . 72

**APPENDIX B:  East Bay Hills Fire Chronology on October 20th** . . . . . . . 73

**APPENDIX C:  Contra Costa County Summary** . . . . . . . . . . . . . . . . . . . . 85

**APPENDIX D:  Strike Teams** . . . . . . . . . . . . . . . . . . . . . . . . . . . . . . . . . . . 88

**APPENDIX E:  Photographs** . . . . . . . . . . . . . . . . . . . . . . . . . . . . . . . . . . . . 92

# The East Bay Fire
# Oakland-Berkeley, California
# October 19-22, 1991

Local Contacts:   P. Lamont Ewell
Director of Fire Services
Assistant Chief Don Matthews
Assistant Chief John Baker
Assistant Chief Andrew Stark
Battalion Chief Neil Honeycutt
Battalion Chief Manual Navarro
Battalion Chief Reginald Garcia
Captain Donald Parker
Oakland Fire Department
1605 Martin Luther King
Oakland, California 94612

Chief Gary Cates
Battalion Chief Paul Burresteros
Berkeley Fire Department

Assistant Chief Bill Cullen
Contra Costa County Fire Protection District

California Department of Forestry and Fire Protection
Governor's Office of Emergency Services

Bill Patterson
FEMA
Region IX

## OVERVIEW OF THE FIRE

The largest dollar fire loss in United States history occurred in the East Bay Hills, within the California cities of Oakland and Berkeley, between October 19 and 22, 1991. Twenty-five lives were lost and more than 3,000 structures were destroyed by a wildland-urban interface fire in one of the most heavily populated metropolitan areas of the North American continent. The fire completely overwhelmed the firefighting forces of the area, consuming everything in its path, and was only stopped when the Diablo wind conditions abated. The wind had threatened to drive the fire across the entire city of Oakland.

The factors that set the stage for this disaster were identified long before the fire occurred, and the potential consequences had been predicted by fire officials. Nevertheless, their warnings went unheeded, and the measures that could have reduced the risks were not implemented. More than one billion dollars in damage resulted from a fire that exceeded the worst expectations in the most concerned fire professionals. It was a fire that demonstrates how natural forces may be beyond the control of human intervention and should cause a renewed look at the risk of wildland-urban interface fire disasters.

Large areas of California are known to be critically vulnerable to wildland-urban interface fires due to the development of urban areas in locations that are subject to extreme fire hazards created by climate, terrain, and natural fuels. Several major fires have occurred over the years, including one in 1970 that involved a large portion of the area burned in this incident. The coastal region was particularly vulnerable in the fall of 1991, after five years of drought, several months with no recorded precipitation, and reduced efforts to control wildland interface hazards due to State and local budget limitations. The key ingredient in this incident was the Diablo wind condition, which combined with the other critical fire risk factors to create an irresistible destructive force.

On the following page is a map of the area of Oakland where the fire occurred. This same map is repeated several times later in the report overlaid with arrows illustrating the direction and development of the fire at various stages. A regional map appears on page 6. This fire was originally labeled as the "Tunnel Fire." It is now being described as the East Bay Hills Fire in most reports.

## SUMMARY OF KEY ISSUES

| Issues | Comments |
| --- | --- |
| Location | Wildland-urban interface area, Oakland-Berkeley Hills, California. |
| Risk Factors | Extreme fire risk created by five year drought, low humidity, and Diablo winds; highly combustible natural fuels, inadequate separation between natural fuels and structures; unregulated use of wood shingles as roof and siding material; steep terrain, homes overhanging hillsides, narrow roads, limited access, limited water supply. |
| Mitigation Efforts | Previous fire experience identified hazards. Risk reduction measures had been studied and recommended for several years, but not implemented. |
| Cause | Strong winds caused rekindle of grass fire from previous day, accelerated by wind. Crews were on scene overhauling when fire erupted. Cause of original fire was undetermined. |
| Response | Largest response ever recorded. Massive mutual aid provided by 440 engine companies and more than 1,500 firefighters. |
| Damage Extent | 3,354 structures destroyed, 1,500 acres, 1.5 billion dollars in damages. |
| Death and Injuries | 25 lives lost, including a battalion chief and police officer; 150 people injured. |
| Incident Command | Shortage of command officers handicapped initial implementation of incident command system (ICS). Multiple commands developed as additional agencies became involved. Unified command implemented after several hours. |
| Communications | Radio channels and communications center overwhelmed by situation. |
| Strategy | Initial attack overwhelmed; crews had to retreat and evacuate residents ahead of fire. Unable to stop advance of fire until wind conditions changed. |

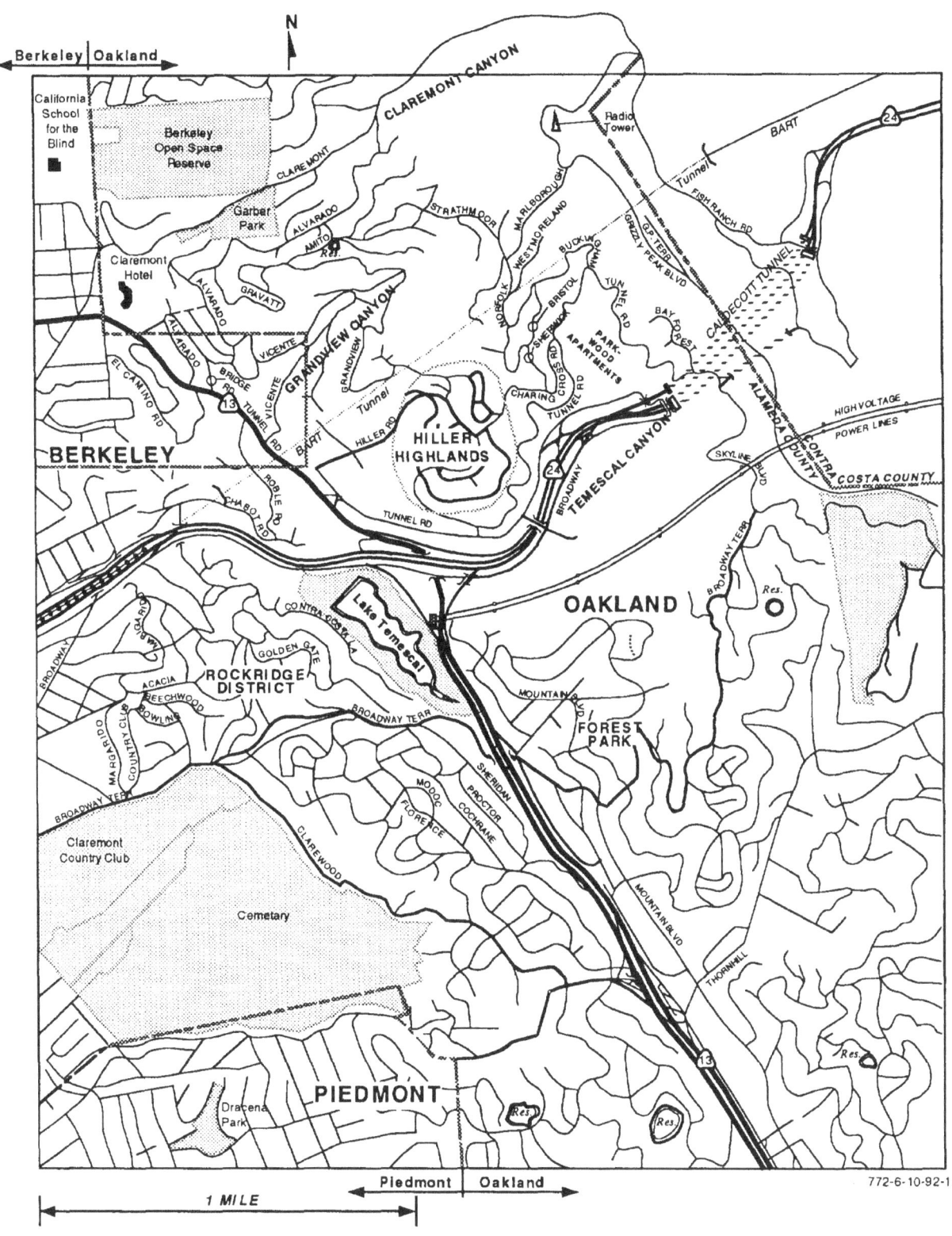

772-6-10-92-1

## LOCATION

The San Francisco Bay area is one the most heavily occupied areas in the United States.  It includes six counties, San Francisco, San Mateo, Santa Clara, Alameda, Contra Costa, and Marin, which surround San Francisco Bay, with a combined population of more than five million.  The metropolitan area includes three major cities:  San Francisco, San Jose, and Oakland, and dozens of smaller communities, many of which are contiguous.

The city of Oakland is situated on the eastern side of San Francisco Bay, directly opposite the city of San Francisco, as seen on the regional map that appears following.  Most of Oakland's 360,000 citizens live on the coastal "flatlands" that extend inland for approximately four miles to the East Bay Hills.  The flatlands are heavily developed with waterfront port installation, the central business district, industrial and warehouse zones, and several residential areas.  Along the eastern edge of the flatlands, the ground begins to rise to a median elevation of approximately 400 feet and it occupied by middle- and upper-class residential neighborhoods.

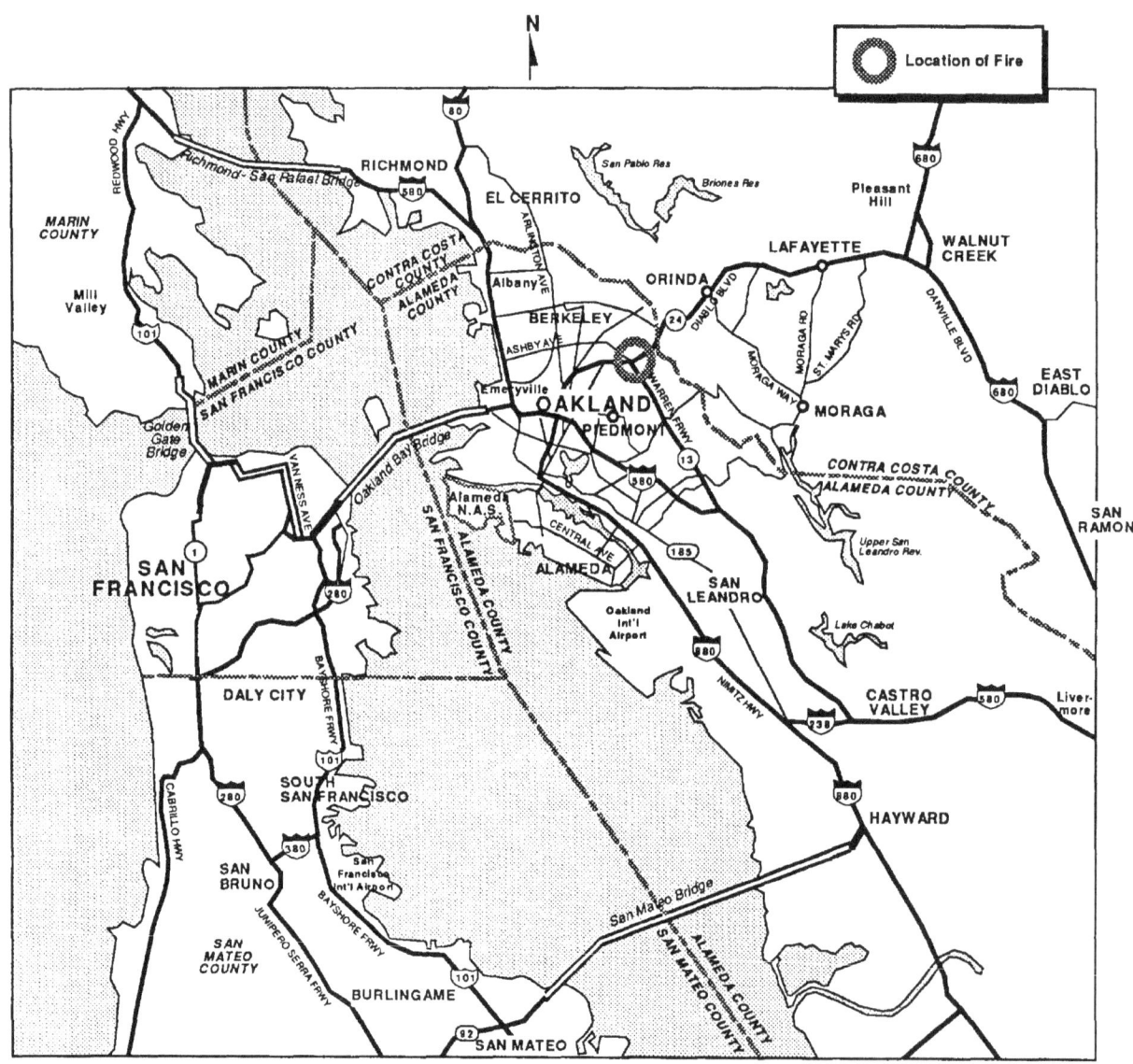

772-10-16-92-1

The terrain then rises abruptly to form a row of hills called the East Bay Hills or the Oakland Hills, with a ridge line approximately 1,300 feet above sea level. The ridge line runs generally in a north-south direction, parallel to the shoreline of San Francisco Bay and approximately five miles inland. The hills separate the coastal flatlands from the inland valleys of Contra Costa County, and the ridge line establishes both the eastern city limits of the city of Oakland and the eastern boundary of Alameda County.

The west face of the hills is heavily developed with expensive residential properties, which are provided with spectacular views of Oakland and San Francisco.

In these hills, Temescal Canyon is one of a series of canyons that open toward the west. Grandview and Claremont Canyons are both north of Temescal Canyon. The canyons separate "fingers" of hills which project up to one mile west of the main ridge line. Temescal Canyon creates a natural path to a narrow point in the hills, providing the shortest distance for a tunnel connection between Oakland and Contra Costa County. This tunnel, the Caldecott Tunnel, links Oakland with the growing communities of Orinda, Moraga, Lafayette, Walnut Creek, Concord, and the San Ramon Valley. It is the only major highway connection in a stretch of more than 20 miles and its triple tubes carry eight lanes of commuter traffic under the hills to the major employment centers of Oakland and San Francisco. A newer tube, approximately one quarter mile north of the Caldecott Tunnel, carries the Bay Area Rapid Transmit System (BART) under the hills.

Highway 24 follows the bottom of Temescal Canyon from the tunnel portals to the mouth of the canyon, where it meets Highway 13 in a Y-shaped interchange, then continues west toward downtown Oakland and the Bay Bridge. Highway 13 carries north-south traffic along the base of the hills, before turning west into the city of Berkeley. Within the Y of the freeway interchange are an electrical substation and a small recreational area surrounding Lake Temescal.

Grizzly Peak Boulevard follows the ridge line, approximately 600 feet above the Caldecott Tunnel, barely within the city of Oakland.

The East Bay Regional Parks District administers several recreation and preserve areas along the hilltops and on the slopes, straddling the county line. Other parts of the hill area belong to the University of California – Berkeley. The southeast corner of the city of Berkeley includes part of Grandview Canyon, just north of Highway 24 on the Oakland side. On the Contra Costa side, the communities of Orinda and Moraga lie at the base of the hills and include residential areas that have been developed on the eastern slope.

## BACKGROUND

### Climatic Conditions

The East Bay Hills have their own micro-climatic conditions, distinct from the areas to the east and west. The "flatlands" have a cool damp coastal climate, influenced by San Francisco Bay and the Pacific Ocean. The prevailing winds push moist air against the Oakland side of the hills and often create gusty winds, while the flatlands atmosphere is calm. The hills block low clouds and moist air coming through the Golden Gate opening from the ocean, keeping the moisture from reaching Contra Costa County. There is often a temperature differential of 50 to 100 between the hills and the flatlands, with the flatlands cooler sometimes and the hills cooler sometimes.

Between 1986 and 1991 most of California experienced drought conditions. This situation was recognized as creating more and more critical fire risk conditions each year. The unprecedented drought was accompanied by an unusual period of freezing weather, in December of 1990, which killed massive quantities of the lighter brush and eucalyptus. Dead fuel accumulated on the ground in many areas and combined with dropped pine needles and other natural debris to create a highly combustible blanket. Due to the fiscal cutbacks, governmental programs to thin these fuels and create fuel breaks were severely curtained, so the fuel load was much greater than normal by the second half of 1991. In addition, no measurable rainfall was recorded during the summer and early fall of 1991.

The coastal areas of southern California are extremely vulnerable to the infamous Santa Ana wind, officially classified as a foehn wind condition. A similar condition occurs in the Oakland area, where it is known as a Diablo (or "Devil") wind. These winds are created when a high pressure weather system is located over the great basin of the inland western States, accompanied by an offshore low pressure system. The high pressure system imports chilled air from the far north, with extremely low moisture content. The interaction of the two pressure systems and their counter-rotational forces creates a wind flow from northeast to southwest, while the pressure differential forces the dry air from high altitudes down to ground level. The result is a strong wind of exceptionally dry air, blowing through the mountain passes and spilling over the coastal hills toward the Pacific Ocean. Increased pressure also heats the air mass (adiabatic compression), which often results in air temperatures of 90 to 100 degrees Fahrenheit at sea level, with less than 10 percent relative humidity and wind velocities of 35 to 70 miles per hour.

Most of the major wildland fires in California have occurred during foehn wind conditions, which occur most frequently between mid-September and late November. The fire protection agencies in California are highly aware of the danger that is created by these wind conditions and use a red flag alert system to warn of extreme fire risk conditions. The Weather Service monitors weather trends in the western States to issue an early warning of impending red flag conditions.[1]

The drought conditions prevailed through October 1991 with warmer than normal temperatures. A Diablo wind condition was predicted for Sunday, October 20, and red flag warnings were issued to wildland fire agencies.

## Vegetation

Most of the native trees on the East Bay Hills were cut during the 1800s to provide wood for railroad ties and lumber for building construction, leaving the hillsides almost bare. In the early part of this century, a major effort was instituted to reforest the hills and several non-native species were imported. Eucalyptus trees from Australia were selected because of their rapid growth and their ability to cover large areas with green trees. Monterey pine trees were also imported form other parts of California.

The west face of the hills receives significantly more moisture than the east face, encouraging the growth of trees and brush on the Oakland side. The Oakland hills are covered with dense growths

---

[1]The red flag alert indicates that prevailing conditions present an extremely high fire danger. The alert is posted at all park and forest service facilities and signifies that severe restrictions are in effect to prevent fires. In many areas camping privileges and other wildland activities are suspended during red flag alerts.

One fire chief commented, "If the Oakland hills had been part of a national park or forest, instead of a residential neighborhood, the area would have been evacuated during the red flag weather conditions."

of trees, supplemented by grasses and thick brush. The east face is exposed to the more arid climate of the inland valleys and is predominantly covered by grasslands and brush.

These particular trees and brush are highly vulnerable to rapid fire spread and release massive amounts of thermal energy when they burn. They also create flying brands, which are easily carried by the wind to start new spot fires ahead of a fire front. The extreme fire hazard of these fuels is greatly exacerbated by the steep terrain and by adverse humidity and wind conditions.

## Land Development on the Hills

The first residential areas in the hills were developed in the early part of this century, after the San Francisco earthquake of 1906, as country homes for the upper class families of the crowded city across the bay. Several large homes were built along the lower slopes in Grandview and Claremont Canyons. The landmark Claremont Hotel was built by a developer to attract potential buyers to the area. By the 1920s, homes for the upper-middle class were being built along the lower slopes and several very large mansions were built on the higher ground.

The major development of Temescal Canyon took place in the 1950's and early 1960's. Narrow switchback roads were built on the canyon slopes and along the ridges, and expensive homes were built to take advantage of the spectacular views.

The roads include several steep grades and hairpin turns and, in many places, are so narrow that two automobiles have difficulty passing. The access for large firefighting vehicles is very limited. The water supply is also limited, particularly when evaluated against the combined fire risk of dense vegetation and large wood-roofed homes on steep slopes. Several water storage tanks are provided at different elevations to serve domestic requirements and hydrants in the area. The water system relies on electric pumps to relay water from the lower storage tanks up to the tanks at higher levels.

The homes on the hillside range in value from 250,000 dollars to several million dollars. The steep slopes of the canyon walls present difficult construction challenges. Many were built on platforms overhanging the canyon walls or with multiple levels stepping down the hillside. Garages, sundecks, or swimming pools were often constructed on the top level, with two or more levels of living area below the level of access from the street. Short bridges were required in many cases to span between the street and the garage entrance. Untreated wood shake or wood shingle roofs were common, and no requirements for fire resistive roof coverings or walls were enforced.

The combination of natural fuels and human-built structures created a critical wildland-urban interface. Many of the structures were completely enveloped in the natural fuels, including the areas below overhangs. The trees were so dense in many places that they created a natural canopy over the roads and no regulations for clear areas between wildlands and structures were enforced. The steep slopes created a natural draw for a fire to spread up to, under, and around the homes.

Two major developments occurred after 1970. One brought Hiller Highland, a neighborhood of 340 densely-built, two-story condominiums and townhouses to the hills. New, wider access roads were built to serve this development, but the heavy overgrowth of trees was maintained. A single connection was provided to Charing Cross Road, one of the older hillside streets.

The second major development was the Parkwood Apartments, a complex of seven buildings, containing 456 apartments, that was built at the bottom of Temescal Canyon, just north of the Caldecott Tunnel entrance. The buildings had three and four stories of wood-frame apartments, built on top of open parking areas, and were terraced into the lower levels of the canyon. This exclusive development had only one access point, controlled by a security gate, connecting to the Highway 24 frontage road.

The Rockridge district, west of Temescal Canyon, features small hills and hollows that were covered with large, older homes, mainly two stories in height and closely spaced. This area was also rich in natural and ornamental trees and brush. Large pine trees were planted around many of the homes, and the mature trees enveloped many of the structures. The entire area featured lush landscaping, and many of the homes had wood shingle roofs, although Spanish tile roofs were also found in the area.

## Burning Characteristics

Fire has been a part of the history of the Oakland-Berkeley Hills area throughout its history. As with many other marine climates, fuel moistures are such that during most periods, fires do not cause dramatic damage but rather help maintain a balance of fuel types and reduce fuel loads. The native flora and fauna had adapted correspondingly with the natural occurrence of fire in the area.

In modern times, the natural fire pattern in the area has been substantially changed. Fire suppression has reduced the natural cycle of fires which normally would have occurred in the area. Without prescribed burning or some other type of fuel reduction, the native vegetation has caused an increased fuel load through the area.

Additionally, the introduction of vegetative species which are not native to the area has dramatically impacted fuel loading. This is particularly true of the introduction of eucalyptus. Fuel accumulations in some areas under eucalyptus plantations have been estimated between 30 and 40 tons per acre. Monterey Pine was also introduced into the area and contributed significantly to the fuel loading.

Eucalyptus was first introduced to the East Bay Hills with extensive planting in the early 1900s. The eucalyptus has a tremendous production of both leaf and bark litter, which is not readily consumed or broken down in the normal decomposition process and leads to the presence of high volumes of fuel.

Additionally, eucalyptus is susceptible to freeze damage, as occurred in 1972, when large numbers of eucalyptus were killed due to an extended period of below freezing temperatures, and again in December of 1990. The dead trees and limbs added a significant amount of dry fuel in the area. Also, eucalyptus sprouts back from the stump and this sprouting after freezing or after logging operations has also increased fuels in some areas.

Fuel loading varied through the fire area and to some degree was dictated by the topography. The northeast portion of the fire area had more wildland fuels, while in the south and western areas, the homes were the major fuels. In effect, the more severe slopes in the north and eastern portions of the fire area required the use of native species. The more moderate slopes and deeper soils in the south and southwest areas allowed for the introduction of more ornamental type species.

**Fuel Distribution** — Natural vegetation types through most of the fire area included some grassland, brushland, mixed broadleaf forest, and eucalyptus and conifer plantations. Species found in the grassland areas include various types of needle grass and perennial bunch grasses. Heavy grazing in the areas in the past resulted in the introduction of various annual grasses, such as wild and slender oats, barleys, soft chess, other bromes, and an array of associated annual and perennial herbs. Additionally, thistles, mustards, and wild oats dominate some of the area. With the discontinuation of grazing, coyote brush has become established throughout the grassland area.

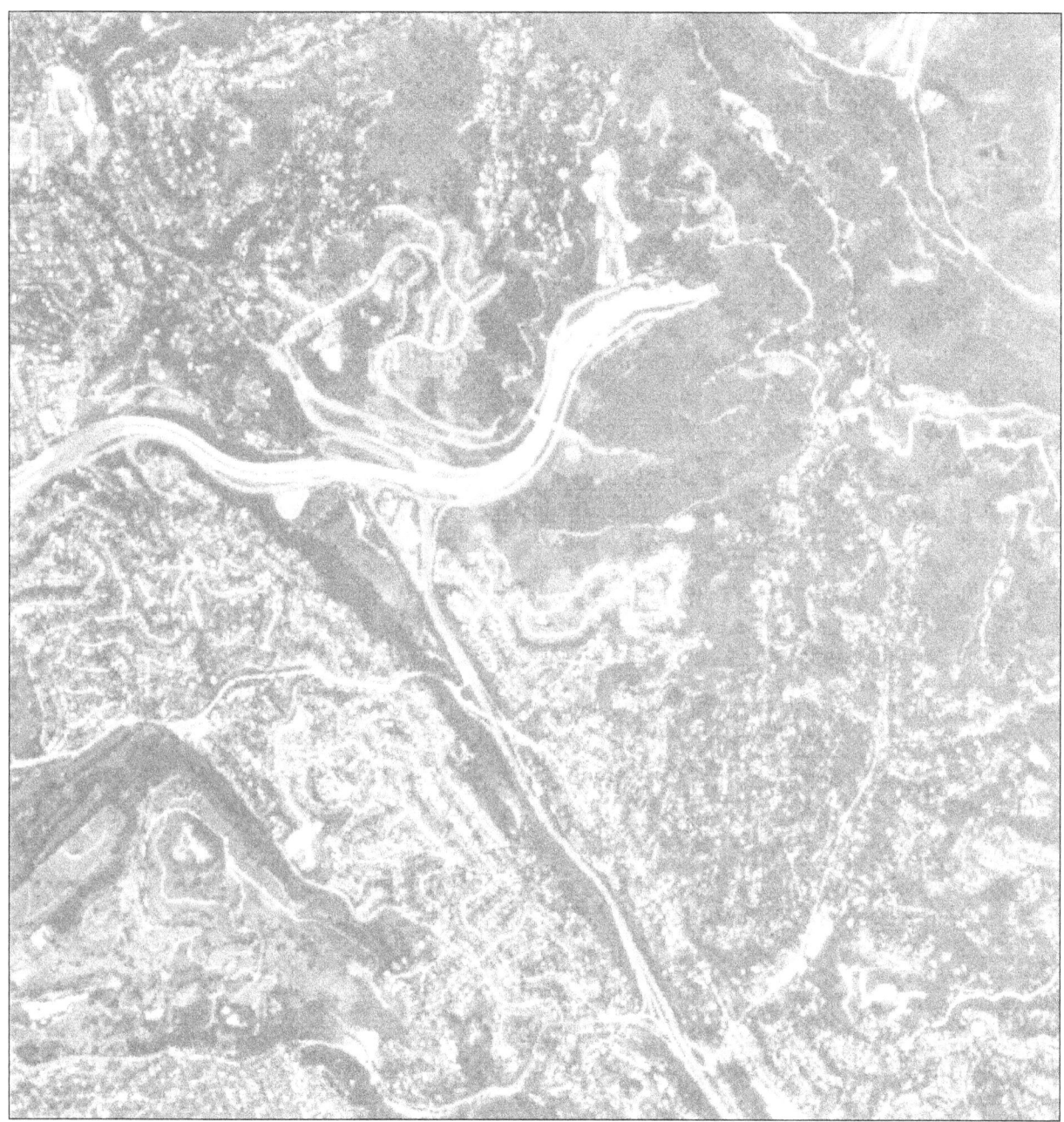

This is an aerial photograph of the East Bay Hills area taken before the fire. By comparing it with the map on page 3 the reader will be able to identify the Hiller Highlands, the Parkwood Apartments, Lake Temescal, the Claremont Hotel, and major roads. A similar photograph, taken after the fire, appears on page 64.

The grasslands would contain the lowest fuel loading of the natural fuels through the area. However, the extended drought may have caused unusual amounts of dead fuel to accumulate before the fire.

The brushland would probably make up a large portion of the available fuel, particularly in the northeastern portion of the fire area. Drier sites would contain such things as silver lupin, California sage, and bush monkey flower. Other species which might be found would include poison oak, coffeeberry, ocean spray, and hazelnut.  Hard chaparral type brushland would include alameda and brittleleaf Manzanita, bush chinquapin, and huckleberry.

Mixed broadleaf forest fuels were located in portions of the fire area, but were probably not significant in the fuel loading.  Species found would have included coast live oak, California Bay, and California Buckeye.  Some Monterey pine and eucalyptus may also have been found scattered through these locations.  Understory species would include poison oak, blackberries, hazelnut, and various herbaceous plants.

Eucalyptus plantations were found at various locations, through the fire area.  Numerous eucalyptus, not specifically in plantations, were also scattered throughout the fire area.

Conifer species found through most of the area would have been Ponderosa Pine or the introduced Monterey Pine.  There would be also some minor locations with Coastal Redwood.   Monterey Pine plantations have been established through portions of the area.

In addition to the native species, various ornamental species have been added throughout the area, particularly in and around the homes.  These fuels appear to be more significant toward the south/southwest of the fire area.

**Fuel Loading** – The heaviest fuel loading would probably have occurred in the untreated eucalyptus stands.  Some estimates indicate fuel loading in these areas from 30 to 50 tons per acre.  Additionally, heavy fuel accumulations would have occurred in the location of the brush lands.

Information indicates that some fuel reduction had been attempted through portions of the area.  Some logging had occurred to thin eucalyptus stands and some attempts had been made to remove eucalyptus from certain areas.  The East Bay Regional Parks District (EBRPD) has used grazing to reduce fuels in some areas.  The amount of fuel reduction conducted by individual homeowners varied greatly.

Various fuel reduction options need to be considered and developed in these areas.  Prescribed burning and limited grazing by animals such as goats are two options that have been considered.  Mechanical methods of fuel reduction could also be utilized.

**Fire Spread** – Fire spread through the majority of the north and eastern portion of the fire was probably aided by the continuity of wildland cover and the rapid burning characteristics species.  Most of the residences and other structures located through this area burned as the result of being exposed to the fire spreading through the continuous wildland cover.  As the fire progressed into the more heavily populated area to the south and west, the structures themselves contributed to spreading the fire, assisted by the wildland and ornamental species.

Entry of fire into the structures through the fire area was closely associated with the adjacent wildland or ornamental plant species.  Many of the homes in the steeper slope areas had overhanging decks with fuel accumulations underneath, allowing fire to spread to the decks and into the structures from below.  It appeared from the location of trees around the homes that fuel accumulations on the roofs probably added to the spread of fire to the roof coverings and under the eaves.  Additionally, fuels in close proximity to the structures proved to be significant, exposing the exteriors to extreme radiant heat loads.  In many cases, the radiant heat caused interior contents inside windows to ignite.

Strong winds and low fuel moistures resulted in an extremely fast spread of fire.  This was increased by the topography and heavy fuel loading.  The extremely rapid rate of spread, coupled with the difficulty of control made a frontal attack to stop the forward progress of the fire almost impossible. Suppression efforts in the initial stages of the fire consequently had to be defensive action in an attempt to control exposures and contain the spread on the flanks of the fire.

Spot fires occurred a quarter of a half mile in front of the fire as the wind carried embers over a large area. Spotting occurred in both wildland fuels and on top of structures. There were also spot fires occurring in the various ornamental species and in yards in the more residential areas of the fire.

### Ignition Sources

The combination of weather, terrain and fuel conditions have made large areas of California extremely vulnerable to wildland fires. The most frequently reported ignition sources are lightning, arcing power lines, human carelessness and incendiarism. Lightning is a high risk factor when it occurs without accompanying rain. The Diablo winds often cause power lines to arc or short out, sometimes providing multiple ignition sources as sparks rain down on the dry grass and brush.

Human carelessness includes unattended open fires and smoking materials as ignition factors. Regulatory measures are often implemented to restrict open burning and to limit or prohibit entry into high risk areas during periods of high risk weather.

Arson has been a major problem in interface areas. Individuals who seek to cause large scale destruction recognize the factors which make California's coastal areas extremely vulnerable to wildland fires. Efforts to warn the public of critical fire danger periods have been identified as an invitation to arsonists to take advantage of the opportunity. A disproportionate number of arson fires have occurred in the most vulnerable areas, during the most critical periods.

The exact cause of the East Bay fire has not been determined. The major fire originated as a rekindle of the fire which occurred and was controlled on the previous day. Several leads were pursued relating to accidental or negligent causes for the original fire, but no definitive determination was made. A number of arson fires occurred in the hills, before and after this fire, including at least one on the same day, but no link has been made to any of those incidents.

## Previous Fires In The East Bay Hills

The East Bay Hills have been the scene of a number of wildland-urban interface fires over the past 70 years. The circumstantial factors surrounding the major incidents have been remarkably similar. The Santa Ana wind condition, preceding periods of unusual dryness, wood shingle roofs, high burn rate natural fuels, lack of separation between the natural fuels and structures, lack of natural fuel controls, poor access, limited water supplies, and difficult terrain have all been recognized as factors in the previous fires.

### History of Major Fires in the Oakland-Berkeley Area

| September 17, 1923 | Berkeley Fire | 640 Structures |
| September 22, 1970 | Fish Canyon Fire (Oakland) | 39 Structures |
| December 14, 1980 | Wildcat Canyon Fire (Berkeley) | 5 Structures |
| October 20, 1991 | East Bay Hills Fire (Oakland) | 3,354 Structures; 25 Lives Lost |

The most remarkable similarities can be seen from comparisons of the 1923 and 1970 fires with the 1991 fire. All three fires originated in the hills and spread into developed areas, pushed by Diablo winds, and each one continued to spread until the wind abated. The 1923 fire originated in Wildcat Canyon, approximately 2-1/2 miles north of Temescal Canyon, and burned from the hills down into the city of Berkeley "flatlands." This fire consumed 640 structures to the north of the University of California Berkeley campus. Wood shingle roofs, the wind, and dry weather were cited as the major factors in this conflagration. Recommendations were made after the fire to limit the use of wood roof coverings and to control the natural fuel conditions in the hills.

The 1970 fire originated on the eastern slope of the hills near Fish Ranch Road, just over the ridge from Temescal Canyon. It spread rapidly up the slope and jumped over Grizzly Peak Boulevard on a front 300 to 400 feet wide. The fire then spread down into Temescal Canyon and subsequently crossed over into the upper parts of Grandview and Claremont Canyons. It was controlled at that point when the wind condition became less severe. The 39 homes that were consumed included virtually all of the homes that existed in the burn area in 1970. All of the homes that were rebuilt in this area were destroyed again in the 1991 fire, along with dozens of additional homes that had been constructed in the intervening years.

The 1970 fire followed virtually the same path as the early stages of the 1991 fire[2] and the losses were attributed to exactly the same factors: wind, weather, natural fuels, lack of separation between structures and natural fuels, unlimited use of wood shingles, terrain, access, and water supply were all identified as major factors in both fires. Investigations that followed the 1970 fire recommended regulatory restrictions to mitigate some of the risk factors, *but the area was permitted to be rebuilt and additional development was allowed to occur without action on the recommendations.*

The 1980 fire also originated in Wildcat Canyon and spread rapidly to involve five homes in the immediate area. The key factors that were identified in this fire were the lack of separation between natural fuels and structures and the unrestricted use of wood shingle roofs. While the wind was a factor in this fire, it was not as strong or persistent as in the other fires, and the fire was successfully contained. This fire underlined the risk factors that are created by the intimate mixture of structures into highly combustible natural fuels and by the use of wood shingle roofs.

## Other Wildland-Urban Interface Fires in California

The 1970 fire occurred during a period of eight days in which three major wildland-urban interface fires occurred in California. During a 60 day fire season in that year, 1,260 fires burned 600,000 acres, destroyed 885 homes, claimed 14 lives, and caused an estimated 233,000,000 million dollars in damage along the California coast.

The Bel-Air Fire, in November of 1961, destroyed 537 structures in the Santa Monica Mountains, in the city of Los Angeles. This fire was preceded by the Laurel Canyon Fire in July of 1959 which destroyed 38 homes. Several major investigation and analysis projects were conducted after these fires and were consistent in their warnings and recommendations. In each of the previous cases the wood roof and separation from natural fuels problems were emphasized.

---

[2] The 1970 fire originated on the east slope of the hills and crossed over the ridge into Temescal Canyon. The 1991 fire originated on the west slope in the area where the 1970 fire began to spread on that side of the hills.

In 1990, the Paint fire in Santa Barbara County involved 430 structures and caused one death. This was the largest loss of structures since the Bel-Air Fire and reminded public officials along the California coast of their extreme vulnerability to interface fires, particularly in view of the ongoing drought conditions. As the drought continued into 1991, fire officials were extremely concerned with the risk of one or more major fires that could have even more devastating results than any of the previous fires, particularly if the Santa Ana winds made their anticipated appearance during the period from September to November.

## Regulatory Efforts

Several regulatory efforts have been made to reduce or control the risk of wildland-urban interface fires in California. Fire officials were very much aware of the critical fire risk factors that were present in the East Bay Hills and had often encouraged measures to mitigate the risks.

Wood shingle roofs were identified as a major fire risk factor well before the Berkeley fire in 1923. Several urban conflagrations in different parts of the United States and Canada were attributed to wood roof coverings in the 1800s and early 1900s. By 1923, many States had placed regulatory restrictions on their use.

A 1959 report by the National Fire Protection Association (NFPA) identified the wood shingle risk factor and encouraged officials in California and Texas (which has a similar problem with wood roofing materials) to take action. This report illustrated the vulnerability of wood shingles to ignition, as well as their tendency to spread flaming brands downwind, starting new spot fires well ahead of a main fire front. This factor was identified in the 1959 fire in Laurel Canyon and the worst predictions were realized in the 1961 Bel-Air Fire. Nevertheless, legislative efforts to restrict the use of wood roofing materials were successfully resisted by the lobbying efforts of the wood products industry, which is a major component of the economy of California. The city of Berkeley passed prohibitive ordinances after the Wildcat Canyon Fire in 1980, but the city council was persuaded to rescind the regulations soon after their adoption.

The parallel concern for maintaining an adequate clearance between natural fuels and structures has also been a subject of regulation in several areas. Following the Bel-Air Fire, most jurisdictions in southern California began to enforce strict requirements to maintain separations between structures and natural fuels, particularly in hillside areas. The State of California adopted regulations under the Public Resources Code which require a 30 foot fire break around all structures in interface areas and require fuel modification within a 100 foot radius. These regulations apply only to State responsibility areas, however, and must be adopted by local jurisdictions to apply to their areas.

Minimum requirements for water supply, roadway widths, and access for fire apparatus have also been adopted for State responsibility areas and in many local jurisdictions. In most cases these regulations cannot be retroactively enforced on existing developments.

In 1982, a Blue Ribbon Committee studied the problems of fire risk in the wildland-urban interface areas of the East Bay Hills and recommended a major effort to institute fire breaks and fuel control measures. The fuel control measures on public lands were partially implemented for a few years, but fell victim to the economic crisis in the years preceding the 1991 fire. Most of the dead fuels resulting from the 1990 freeze were not cleared during 1991.

The Hazard Mitigation Report, prepared under the aegis of FEMA as a result of the Presidential Disaster Declaration for the East Bay Fire, identified several recommendations to reduce the risk of future fires in the area. The recommendations include requirements that would address fuel management, separations, roofing materials, roadways, and water supplies. These points are all recommended for implementation as the fire area is rebuilt and for application to other areas when feasible.

There has been considerable public pressure, both for and against the adoption of local ordinances or regulations that would change the character of the East Bay Hills. Deed restrictions in some areas prohibit the removal of trees or alteration to the natural ground cover. Proponents of safety have demanded a full review of all regulations before any permits are issued to rebuild.

Many homeowners expressed their desire to rebuild "exactly the way it was," without restrictions on construction materials or fuel separations. The vocal residents wanted to restore the character of the area without governmental restrictions. The same residents expressed fears that widened roads would bring unwanted traffic to the area and demanded compensation for any property taken to make improvements.

## Fire Protection Agencies

The city of Oakland is protected by a 480 member career fire department and operates a total of 23 engine companies, seven truck companies, a hazardous materials (Hazmat) unit, a fireboat, and a light duty rescue unit. Several of the engine companies are assigned small 4-wheel drive brush trucks, known as "patrols" that are used for grass and brush fires and are staffed by the engine crew when needed. An engine company will usually take both vehicles when responding to a brush or grass fire. Most of the engine and ladder companies operate with four crew members, although three member crews are not unusual on engine companies, depending on the available personnel.

The Oakland Fire Department has been severely impacted by budget restrictions during the past decade and has lost approximately 40 percent of its on-duty staffing. At least 10 companies were discontinued and the remaining companies operate with reduced staffing. The hardest hit area was the command level, which was reduced from an assistant chief and five battalion chiefs on each shift to an assistant chief and two battalion chiefs. The assistant chief now serves as both the shift commander and as battalion 2. The chief's operator (battalion aide) positions were also cut, leaving the three on-duty command officers with no support staff.

The city of Berkeley operates seven engine companies and two truck companies, under the supervision of a battalion chief on each shift. Berkeley also reduced its on-duty staffing level by approximately 30 percent as a result of budget limitations. The 94 member department also provides advanced life support ambulances and a Hazmat unit.

Both Oakland and Berkeley are located in Alameda County and participate in the Alameda County Fire Mutual Aid Plan. Oakland is the coordinating department for mutual aid within the north zone of Alameda County, which includes 11 agencies. The Lawrence Livermore Laboratories Fire Department is the central coordinating agency for all three zones (North, South, and East), which includes 29 different fire protection jurisdictions.

The Contra Costa County Fire Protection District is responsible for protecting a large unincorporated area in Contra Costa County, as well as the incorporated communities of Lafayette, Walnut Creek, Concord, Clayton, Pleasant Hill, and Martinez. The Contra Costa County Communications Center (ConFire) dispatches the 20 stations that are operated by the Consolidated Fire District, provides communications for the Orinda and Moraga Fire Departments and serves as the central mutual aid coordination point for the entire county. The remainder of Contra Costa is served by a mixture of 17 municipal fire departments and independent fire districts.

The California Department of Forestry and Fire Protection (CDF) has the primary responsibility for fire protection on State owned and State administered lands. It is primarily staffed and equipped as a wildland fire protection agency, although it provides structural fire protection for some counties and local jurisdictions on a contractual basis. In the San Francisco Bay area, CDF operates wildland engine companies, hand crews, bulldozers, and helitack units. (A helitack unit is a helicopter carrying a lo-member hand crew and capable of making water drops on fires.) Tanker aircraft are also deployed to protect State responsibility areas. The regional headquarters is located at the Santa Clara Ranger Station in Morgan Hills, near San Jose.

The California Office of Emergency Services (OES) is a State government executive level agency that is charged with disaster preparedness and coordination. OES coordinates the Statewide mutual aid system for fires and other major emergencies. The entire San Francisco Bay area is located within OES Region II, which is coordinated from the CDF office in Santa Rosa. Among the resources of OES are a fleet of OES-owned engines that are assigned to local fire departments. These engines are subject to call-up for major incidents and are then staffed by the fire departments where they are assigned. Several of the Bay Area fire departments have OES engines in their stations and are prepared to respond with them.

Most of the Bay area is protected by career fire departments, although some of the smaller jurisdictions have volunteer reserves for backup. Alameda County has a small reserve force, organized under OES, which responded to the East Bay Hills fire. In addition to the municipal fire departments, fire districts, and CDF, there are also several military installations and independent facilities, such as the Lawrence Livermore Laboratory, the Lawrence Berkeley Laboratory, and the East Bay Regional Parks District (EBRPD) that have their own career fire departments and participate in the mutual aid system.

## HOW THE FIRE STARTED

### Point of Origin

The East Bay Hills fire originated on the steep slope at the very end of Temescal Canyon. The canyon turns north from the portals of the Caldecott Tunnel, forming a V-shape that leads directly to Grizzly Peak, the highest point in the area at almost 1,500 feet. Gwin Tank, which is part of the East Bay Municipal Utilities District water system, sits at the top of this slope, near the intersection of Marlborough Terrace and Grizzly Peak Boulevard. The hilltop is occupied by a radio tower, which is used as a transmitter site for a radio station and for some public safety radio channels. The closest Oakland Fire Department companies have an approximate response time of 10 minutes to this location due to the steep hills and narrow roads.

### Saturday, October 19th

On Saturday, October 19, 1991, the weather was warm, clear, and dry, with no appreciable wind. At 1212 hours a brush fire was reported on the hillside above 7151 Buckingham Boulevard, on the end slope of Temescal Canyon. This is one of the steepest parts of the canyon, with a drop of approximately 450 feet between Marlborough Terrace and Buckingham Boulevard. The vegetation on the slope was mostly grass, with some brush and a few trees. The slope directly above the fire was too steep to build on, but there were structural exposures on Westmoorland Drive and Marlborough Terrace, a few hundred feet west of the fire. Additional structures were exposed below the fire, on Buckingham Boulevard.

The actual source of the ignition has not been determined. The fire originated on the slope behind a house on Buckingham and spread rapidly up the hill. In the calm air, the fire spread was directly up the slope and was visible for miles.

The first alarm assignment included three engine companies and the assistant chief. Another battalion chief was in the area and also responded on the first alarm. A second alarm was requested at 1219 hours, followed by the third alarm at 1221, and a fourth alarm three minutes later. Oakland finally struck a fifth alarm for this fire at 1248 hours. This brought a total response of 12 engine companies and two ladder companies from Oakland and two engine companies from the city of Berkeley, as well as three engines and four patrol units from the EBRPD. Companies from Berkeley, Piedmont, Alameda, and San Leandro covered Oakland stations during the fire.

Oakland notified the California Department of Forestry and Fire Protection (CDF) of a fire in the "threat zone," which indicates that the fire could spread into CDF jurisdiction. CDF responded with a first alarm assignment of four engine companies, one helitack unit, one private (contract) helicopter, and a battalion chief. The CDF engine companies were not used, but the helicopters were used to drop water on the fire to help stop its spread and then to quench hot spots on the steep slope. The EBRPD's helicopter, Eagle 5, was used for aerial reconnaissance.

The Oakland Fire Department's command post vehicle was positioned at the off-ramp from Highway 24, just west of the Caldecott Tunnel entrance. This location provided an excellent view of the burning slope and the surrounding canyon. Command officers established divisions on both flanks and at the top of the hill to supervise operations.

The fire was attacked from the lower side by companies on Buckingham and Westmoorland and from above by companies on Marlborough Terrace. The strategic plan was to cut off any potential spread on the flanks and then to squeeze in on the head of the fire to stop the uphill spread. The tactics were successful and the fire was declared under control at 1339 hours. The fire area was limited to two acres, with no structural involvement, and it was stopped on the uphill slope, before reaching the top of the hill.

Overhaul of the fire took several hours and was complicated by the steep slope. While the helicopters dropped buckets full of water on the visible hot spots, hand crews worked to create a secure perimeter around the entire fire area. The hand crews worked the area until darkness made it too dangerous to work on the slope, and the last company did not leave the scene until 1841 hours. Before leaving, a battalion chief checked the area for any visible signs of hot spots and directed the companies to leave the overhaul hoselines in place on the slope. During the night, companies returned to the scene to look for hot spots, and no signs of smoke or flames were observed.

Speaking to reporters after the fire had been controlled, Oakland fire officers noted the tremendous fire potential that was present in the hills after five years of drought, several months with no rain, excessive quantities of dead brush caused by the previous winter's freeze, and the lack of clearance between hillside homes and the natural fuels. They noted that the only factor that was not present that day was the wind – on a windy day the same fire could have become a disaster.

## SUNDAY, OCTOBER 20TH

### The Restart

Sunday morning, October 20, brought the classic Diablo wind conditions to the Oakland area. The weather was still calm on the flatlands as the Oakland Fire Department changed shifts, between 0700 and 0800 hours, but the winds were already picking up in the hills. The battalion chief who had conducted the final check of the fire area the previous evening noticed the weather conditions as soon as he arrived at his home in the hills that morning. He called the on-duty assistant chief (BC2) to make him aware of the situation.

The assistant chief, who had also worked the previous day's fire, recognized the danger and directed two engine companies to check the burn area. Engines 19 and 24 met at the top of the hill, in the area of 7185 Marlborough Terrace, at 0850 hours. They noted a few hot spots, inside the fire line, under some pine trees on the north flank of the burn area and in the upper portion near Gwin Tank.

Engine 19 asked Oakland Fire Communications to have a unit from the EBRPD Fire Department respond to pick up their hose, which had been left on the hillside overnight. Oakland Fire Communications contacted EBRPD and was advised that there would be no one available to respond until their day shift personnel reported for duty.

The assistant chief decided to survey the situation and, while en route, advised Oakland Fire Communications to activate Patrol 28 with overtime personnel. This unit would be assigned to provide a quick response capability in the hill area, due to the windy conditions. Arriving at the top of the hill, at 0913, he advised Oakland Fire Communications to again call EBRPD and request their assistance in overhauling several hot spots that were flaring-up on the hillside. At 0916 he requested an additional engine company and Engine 16 was dispatched.

The wind was continuing to cause flare-ups in the burn area and the assistant chief was concerned with the extreme fire risk conditions. He advised Oakland Fire Communications of the risk factor at 0926 and, at 0929, assumed command of the situation from Marlborough Terrace. At that time several small flare-ups were showing within the burn area.

At the same time, EBRPD was dispatching units to the scene, having received the message that their hose was in danger of burning if it was not soon picked up. The first EBRPD engine arrived at 0932 hours and advised the EBRPD communications center that the situation "seems to be OK" and that the other responding units could "come in Code 2" (without lights and siren).

The flare-ups were controlled by 0945 hours. The assistant chief advised Oakland Fire Communications that the situation was under control and that E24 would be in command of continuing overhaul. As he was leaving the scene, the assistant chief contacted battalions 3 and 4 and directed each one to assign an engine company to patrol the hills, due to the hazardous weather conditions. Engines 27 and 4, from the flatland area of the city, were assigned to this duty. At approximately the same time, the EBRPD fire chief, who was not aware of the flare-ups in Temescal Canyon, contacted his communications center and advised them to increase the staffing at the EBRPD fire stations, due to the weather conditions.

At 0959 hours, the assistant chief questioned the assignment of an engine company to a special assignment at the Training Academy on that particular morning, noting "we have the most critical fire conditions in five years." The assignment was canceled to keep the engine company in service.

Additional EBRPD units and personnel arrived and assisted the Oakland personnel who were over-hauling hotspots on the hillside. Most of the overhaul was conducted with hand tools to root out fire that had burrowed into the thick duff and roots. The hoselines that had been left on the hill were repositioned to cover the perimeter, as a precautionary measure.

The Oakland and EBRPD units were having difficulty coordinating their efforts, since each agency's units were on their own radio channels. Passing messages via the dispatchers, then by telephone from one communications center to the other, proved to be a problem. The units from both agencies were directed to use the "White" (mutual aid) channel to communicate directly.

By 1029 the situation appeared to be well in hand and E24 advised that E19 and E16 would be returning to quarters, while E24 would stay on the scene with the EBRPD personnel. The EBRPD personnel were still working a hot spot on the west flank of the fire, near the bottom of the burn area, which was emitting a significant amount of smoke. At 1035 Oakland Fire Communications advised E24 that they had received a call from a citizen at 7290 Marlborough Terrace reporting a hot spot on the hill. Engine 24 replied that the EBRPD crews were working on that spot and had advised that they could handle it. An EBRPD unit was directed to pull into the driveway at 7151 Buckingham and to extend a line up the hill to cover this flare-up. Radio traffic indicates that the personnel were confident that they could handle the situation and that the flare-ups were all within the safe burn perimeter of the previous day's fire.

At 1040 hours, Oakland Fire Communications transmitted a first alarm for a reported grass fire on Campus Drive, in the hills approximately five miles south of Temescal Canyon. The first arriving unit was Engine 27, which had been assigned to patrol the hills, and the fire was handled by one company.

At 1041 hours, E24 left the scene, leaving E19 in command of the overhaul operation. Engine 16 was already en route back to quarters. The decision to have E19 stay at the scene, instead of E24, was made between the two company officers. At this time the wind was continuing to fan minor flare-ups, but there appeared to be sufficient personnel on the scene to handle the situation with E19 and the EBRPD crews.

## Major Flare-Up

Between 1040 and 1050 hours, the wind velocity increased and several additional flare-ups were observed; the crews were kept busy moving up and down the steep slope to cover them. The EBRPD officer contacted his communications center, reporting numerous rekindles, and requested another EBRPD engine company to respond "Code 3." The EBRPD helicopter, Eagle 5, was also requested to provide a better vantage point to direct operations.

Between 1050 and 1057 hours the winds continued to increase and caused more flare-ups. Engine 19 contacted Oakland Fire Communications to report that one of the flare-ups was in a new loca-tion and there was "pretty good smoke showing" from it. The assistant chief, who had returned to downtown Oakland, ordered E24 to return to the scene. The lieutenant of E19 advised the assistant chief that the smoke was coming from a previously unburned area on the flank of the previous day's fire, but the situation was still under control.

The radio traffic indicates that there was difficulty making radio contact between the Oakland and EBRPD personnel on the hill at this time. The lieutenant of Engine 19 was concerned that some of the EBRPD personnel could be in dangerous positions as several new flare-ups occurred in rapid succession.

Very suddenly, the fire flared up in an unburned area on the lower east flank of the burn area. Burning embers had been carried from one of the hot spots into a patch of timber dry brush. At 1058 hours, E19 called for a "full box alarm" to respond to Gwin Tank. Oakland Fire Communications dispatched E24, E28, and BC2 to assist E19. Engine 4, which was assigned to hill patrol, also responded.

One minute later, Engine 19 requested CDF assistance and, at 1102 hours, requested a second alarm and police assistance for traffic and crowd control. The second alarm units were directed to respond to 7140 Marlborough Terrace, at the top of the hill. At 1104 hours, the assistant chief, who could see the smoke from several miles away, called for a third alarm and also directed Oakland Fire Communications to request mutual aid from CDF. He specified to advise CDF that this was "another fire in the Threat Zone." The Oakland area map presented previously is repeated on the following page showing the location of the restart and its initial spread.

The fire was spreading rapidly uphill, and the strong wind coming over the ridge was pushing the flames out to both flanks at the same time. (A television news crew, who were in the area following up on the previous fire, videotaped the rapid growth of the new fire from a minor flare-up in a growth of brush to a mass of flame, spewing flaming embers on new fuel and igniting new outbreaks at a rapid pace.) The situation changed from offensive to defensive almost instantaneously as the fire raged out of control. Within the first few minutes, E19 reported that the fire was crowning in the trees at the top of the hill, and a spot fire was reported in the area of Norfolk and Marlborough.

The assistant chief drove through the Caldecott Tunnel, noting the heavy smoke and flame on the hillside high above and to the left of the tunnel entrance, and turned up Fish Ranch Road to come up the back side of the hills. As he arrived at Gwin Tank, he could see that the fire was already well beyond the size of the previous day's fire and spreading into heavier fuels. The first structure was becoming involved on Buckingham. He assumed command of the incident and assigned battalion 4 as "Division A" to supervise operations from below on Buckingham Boulevard, while he assessed the situation from the top of the hill. He requested a fourth alarm at 1115 hours.

Engine 24 had come in on the lower side of the fire and reported that they thought they had the fire cut off to the west from 7140 Buckingham. Engine 19, with a vantage point above the fire, could see that left flank of the fire had jumped, possibly from a flaming brand, and started a new run uphill and to the west, above the houses on Buckingham. This created an immediate threat to houses at the end of Westmoorland Drive and at the top of the hill on Marlborough Terrace.

Engine 19 advised the incident commander that there were two distinct fire fronts, moving laterally in both directions from the area of origin. The wind coming over the ridge was meeting the fire spreading up the slope and splitting it into two flame fronts.

An off-duty assistant chief, who was in the area, had responded and was at the bottom of the hill. Communicating with the on-duty assistant chief, they decided to set up the command post vehicle at the Highway 24 off-ramp, in the same location as the previous day. The off-duty chief became the incident commander, while the on-duty chief became the operations officer.

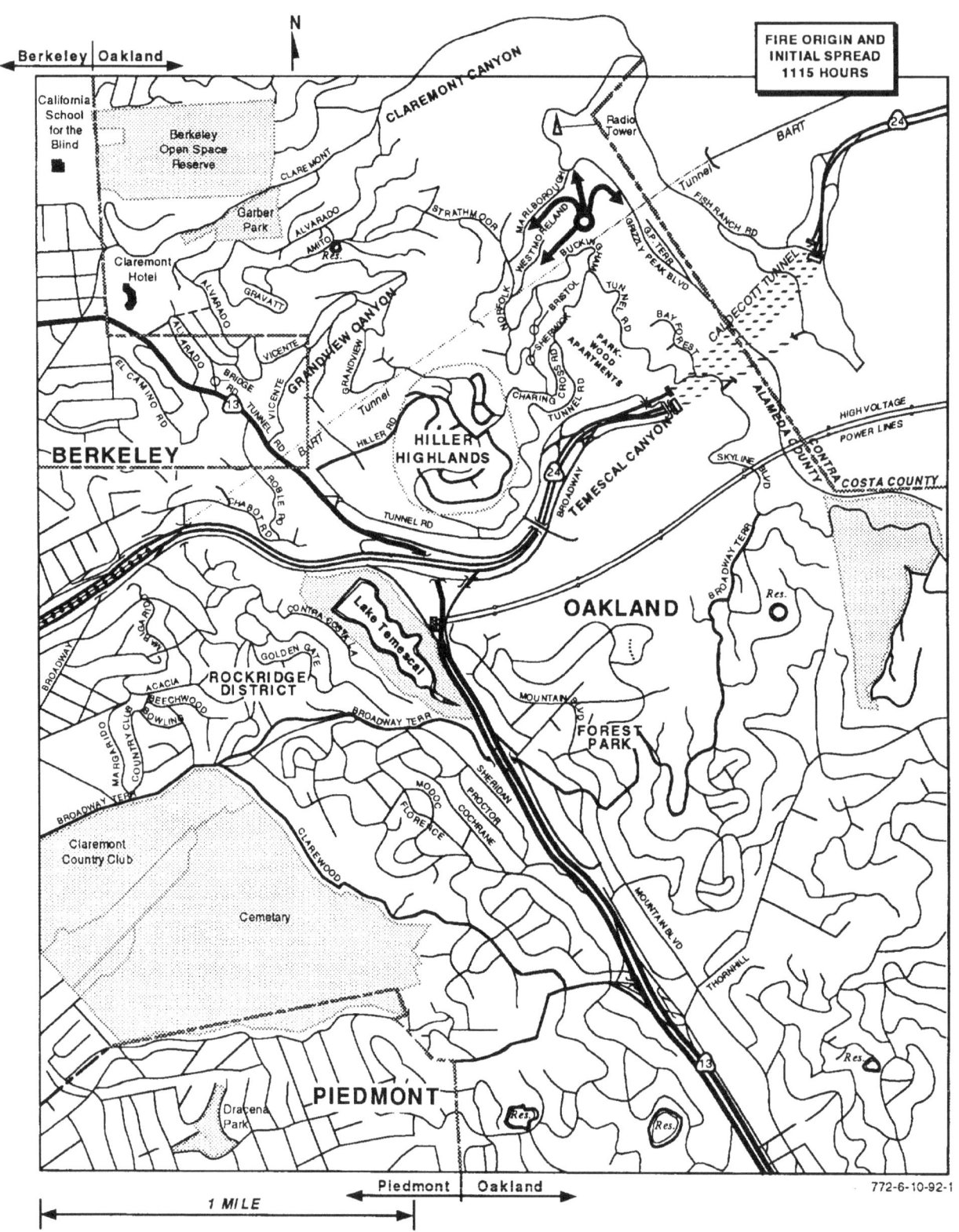

Ladder 1, a tractor-trailer unit, was responding to Marlborough Terrace on the second alarm, when the assistant chief saw the vehicle coming up the hill via Hiller Road. He redirected L1 to meet the command vehicle on Highway 24. The large truck took several minutes to descend via Charing Cross Road to Tunnel Road, with their steep slopes and narrow switchbacks. By the time they reached the bottom of the hill, the fire was spreading to the area they had just come through.

| Situation Status at 1120 Hours | |
| --- | --- |
| *Fire:* | Rapid uphill and lateral spread on both flanks in steep terrain. Moving through fast burning fuels toward residential areas. |
| *Resources:* | Oakland fourth alarm, assigned, assisted by EBRPD, and mutual aid requested from CDF. |
| *Strategy:* | Companies attempting to set-up ahead of fire to protect exposures and hold flanks. |

An assessment of the situation at this point indicates that the rapid **fire spread, combined with very limited access, was beyond the capability of conventional firefighting forces.** To reach the fire, companies had only a few very steep and narrow roads, while the fire had the advantages of weather conditions, terrain, and natural fuels. The wind coming over the end of the canyon and down the slope was splitting the fire to both sides, pushing it directly toward inhabited areas on both flanks. The force of the wind was so strong that master streams were unable to reach the flames. It was clear to the operations officer that rapid intervention with firefighting aircraft was needed to have any hope of stopping the spread of this fire.

At 1119 hours, CDF Morgan Hill was dispatching its closest available helitack unit, Copter 106, to Oakland. The CDF dispatcher had received the initial mutual aid request from Oakland, but when he called back to obtain additional details, the call was placed on hold for several minutes. Copter 106 normally has a 15 to 20 minutes flying time to the fire scene; on this day the strong headwind almost doubled the flying time.

## Communications Problems

Radio communications was a major problem from the outset. The Oakland Fire Department's primary radio channel (Ch2) was overwhelmed with traffic as companies tried to report their approach and request instructions from command officers, report worsening fire conditions, request assistance, and describe their actions. The command officers tried to contact units and make assignments, but the radio traffic was so overwhelming that most messages went without acknowledgements and many were never heard. The only alternate frequency that was available for the command officers to communicate with each other was Oakland Channel 1, and all other Oakland incidents were being handled on that channel, including a structure fire.

Companies were deployed above and below the fire and on both flanks. Without effective radio communications, it was impossible to direct or keep track of them or to maintain any awareness of fire conditions in different areas.

In the Oakland Fire Communications Center, the situation was also out of control. The incoming telephone lines rang continuously, with one caller after another reporting the fire, requesting the fire department to come to a particular address, asking if residents should evacuate, and telling the dispatchers to send more fire trucks to different locations. The news media were calling for information. The radio was so jammed with traffic that it was difficult to hear and respond to the messages that were directed to the communications center. When command officers asked the communications center to do something, the dispatchers were so overwhelmed that several minutes would elapse before it could be done. With so many telephone calls coming in, it was almost impossible to make a call out from the communications center.

Under these conditions, there were several miscommunications and delays in processing information and requests. Automatic notifications had to be made on the multiple alarms, and a recall of off-duty personnel had to be initiated. Mutual aid requests had to be processed. Calls to other agencies were delayed, and callbacks from agencies requesting additional information went unanswered or were put on hold. Additional experienced personnel arrived within the first 15 to 20 minutes to assist in the communications center, but the overload condition persisted for several hours.

## Critical Period

Between 1119 and 1125 hours the situation became even worse:

- Division A (BC4) reported that the fire was spreading rapidly uphill on the west (left) flank and that he would be moving up to Marlborough Terrace to try to protect homes at the top of the hill.

- The incident commander reported that the smoke was so heavy coming down the canyon that he could not see where the fire was or which way it was moving from the command post at the bottom of the hill.

- The fire was spreading south and east on the right flank toward a cluster of homes on Grizzly Peak Terrace. Engine 6 and Truck 15 were assigned to try to protect that exposure.

- The operations officer reported the need for an additional command officer at the top of the hill to cover the right flank. None was available.

- Engine 19 reported that the fire spread on the left flank was lateral, toward the homes on Norfolk Drive.

- Division A reported that he had at least one structure involved, and the fire was "going to jump Buckingham any minute."

The incident commander called for the fifth and sixth alarms at 1120 hours. The operations officer requested the sixth alarm companies to respond to Bay Forest Drive and Tunnel Road, directly above the entrance to the Caldecott Tunnel, where a group of homes were directly in the path of the flames advancing on the right flank. He added a request for five mutual aid engine companies to respond up the back side of the hills to stage at Fish Ranch Road and Grizzly Peak Boulevard.

By 1130 hours the fire was moving so fast on the left flank that companies were abandoning their offensive positions and retreating in search of safe areas to protect structures or to make a stand. Several houses were burning on Buckingham and Westmoorland, and the homes on Marlborough Terrace were in imminent danger as the fire continued to spread west along the face of the canyon and up the slope. The spread of the fire by 1130 is illustrated on the following page.

At 1133 hours, Division A radioed "probably can't hold – it's coming over – we are abandoning task!" The operations officer replied with a warning to be extremely careful: "Don't get anybody killed!" Two minutes later Division A radioed "We're evacuating Buckingham. The fire went over both sides of us!"

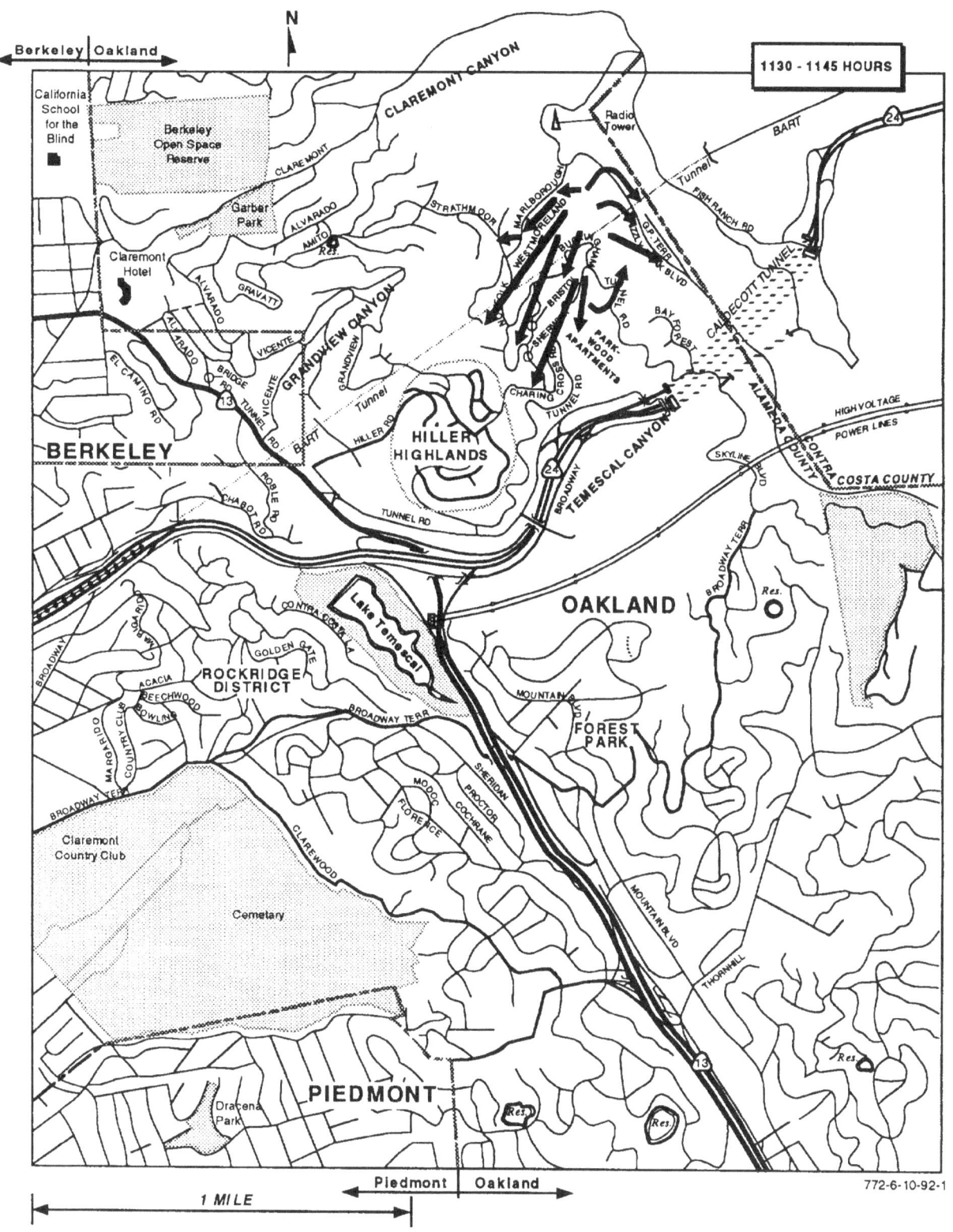

772-6-10-92-1

| Situation Status at 1130 Hours | |
|---|---|
| *Fire:* | Left flank involving structures on Buckingham and Westmoorland, threatening Marlborough Terrace above. Right flank rapidly approaching homes on Grizzly Peak Terrace, Bay Forest Road, and Tunnel Road. Fire threatening to top hill and spread into Contra Costa County. |
| *Resources:* | All available Oakland resources committed or en route. CDF units en route. Mutual aid being requested from Contra Costa County and North Zone of Alameda County. |
| *Strategy:* | Companies on left flank retreating, attempting to evacuate immediately threatened areas. Probably cannot stop fire until resources can be assembled to make a stand ahead of the fire. Companies on right flank trying to assemble adequate resources before the fire reaches structures. |

## Totally Out of Control

At 1133 hours on Sunday, the incident commander instructed Oakland Fire Communications to request five strike teams from Alameda County to stage at Hiller and Tunnel Roads, a half mile ahead of the fire on the left flank. He reported that the fire was totally out of control and moving on several fronts, involving more than 100 acres of trees, brush, and houses. He also requested the response of Pacific Gas and Electric because of numerous power lines that were falling as their poles burned.

This request was intended to provide the resources to make a stand to hold the fire inside Temescal Canyon, trying to prevent its spread into the Hiller Highlands development or over the top to Grandview Canyon. Hiller Road is the only wide access road to the hillside, and Hiller Highlands provided the best opportunity to make a stand ahead of the fire that was moving quickly along the north face of the canyon.

The fire not only burned up the slope and outward on both flanks, but the wind coming over the ridge pushed flames, smoke, and burning brands back down into the canyon. At 1134 hours it became evident that a fire front was moving down the canyon toward the command post. At the bottom of the slope, more than 200 feet below Buckingham Boulevard, the Parkwood Apartments suddenly became an exposure. Highway patrol officers had already closed the Caldecott Tunnel and were trying to clear the backed-up traffic from Highway 24 by sending cars back toward Highway 13. Oakland police officers were sent to warn the apartment residents to evacuate and found some residents already driving out through the narrow security gate. As more residents heard the warnings, the single exit road from the complex became clogged with cars and pedestrians.

The incident commander's next request, at 1135, was for the Oakland Police to send as many officers as possible to help with evacuations, beginning with the 7100 block of Marlborough Terrace.

The incident commander sent the only available companies to try to evacuate and protect the exposed apartment complex. At 1137 hours Engine 1, Truck 1, and Truck 3 tried to enter and work their way toward the rear of the complex, but they encountered backed-up traffic with more than 1,000 residents trying to escape on the private roadways. The plume of superheated gases and the shower of burning brands were beginning to ignite exposed wood surfaces on the upper levels of the buildings at the rear of the complex. Trees among the buildings also ignited, adding to the exposure problem.

Engine 1 worked its way back to a position to operate its elevating master stream, laying a five-inch supply line, but exposures were igniting rapidly and the position in front of the fire became untenable. The crews committed themselves to interior search and rescue, going in after residents who were reported to still be inside the three- and four-story buildings. The incident commander located another available company, Engine 16, and sent them to assist in the apartment complex at 1141 hours. While this was happening, Engine 6 was reporting that the fire was spreading south from Grizzly Peak Terrace, 600 feet directly above Parkwood.

## Mutual Aid Begins to Arrive

The first outside assistance to arrive came up Fish Ranch Road to the back side of the hills. A first alarm assignment, consisting of Orinda, Moraga, and Contra Costa County units, had been dispatched by ConFire at 1135 hours, on a report from a citizen of fire on the hilltop above the Caldecott Tunnel. The Orinda battalion chief (BC45) requested a second alarm at 1143 hours, when he reached the top of the hill and saw the magnitude of the fire on the Oakland side. It was impossible for BC45 to make contact with the Oakland incident commander by radio due to the extremely heavy radio traffic, so the Contra Costa companies went to work with the Oakland companies, trying to protect exposed houses and to stop the fire from topping the hills along Marlborough Terrace and Grizzly Peak Boulevard.

Two minutes after requesting the second alarm, BC45 advised ConFire that Oakland was requesting all available units. Learning that Oakland had established a command post at the bottom of the hill, BC45 assigned a captain as "Division A," to direct the Contra Costa units, while he drove to the command post location.

ConFire had already been contacted by the Alameda County Mutual Aid Coordinator, asking for one strike team and, when the ConFire dispatcher telephoned Oakland Fire Communications for routing instructions, the request was increased to two strike teams. Contra Costa County was also working a multiple alarm brush fire in Franklin Canyon, six miles northeast of this fire. ConFire began to call-up additional strike teams from the remaining resources within Contra Costa County to respond to Oakland.

> **"It's hard to get organized and run for your life at the same time!"**
> *(An Oakland command officer)*

As the fire ignited structures and began to jump the street along Marlborough Terrace, firefighters had to abandon their defensive strategy and evacuate. Companies had already been forced out of Buckingham and Westmoorland, back to Norfolk Road. The firefighting forces were in complete disarray. Streets were blocked by flames and live power lines were falling from burning poles.

### Situation Status at 1145 Hours

| | |
|---|---|
| *Fire:* | Completely out of control, moving west along the north side of Temescal Canyon (left flank) and involving numerous structures. Also involving structures on the upper (Grizzly Peak Terrace) and lower (Bay Forest) right flanks. Spreading rapidly through Parkwood Apartments. Hiller Highlands is in direct path of the fire. |
| *Resources:* | Most Oakland units evacuating or seeking refuge. Contra Costa County second alarm assignment arriving at Fish Ranch Road. Five strike teams requested from Alameda County. Only one helitack available from CDF; all air tankers committed to other fires. |
| *Strategy:* | Companies evacuating residents and abandoning positions. Life safety is only concern until additional resources arrive and can be assembled ahead of fire fronts. |

## Lives Saved And Lost

Most of the fatalities occurred between 1130 and 1200 hours as the fire spread across the north face of Temescal Canyon, involving all of the structures on Buckingham, Westmoorland, Marlborough, Norfolk, Sherwick, Bristol, Charing Cross, and Tunnel Roads. The spread of the fire by 1200 is shown on the following page. Police officers and firefighters tried to evacuate the area as wind-blown brands and embers ignited more and more spot fires ahead of the rapidly moving fire front. Police cars cruised the streets with sirens wailing, and officers used their public announcement speakers to warn residents to evacuate.

Residents who had been standing in front of their homes moments before, watching a fire that was two blocks away, were suddenly piling belongings, children, and pets into their cars. The steep narrow streets, now obscured by swirling smoke, were suddenly clogged with cars as falling power lines and flaming brands ignited spot fires, adding to the confusion. Some of the narrow roads were blocked by collisions as panic stricken residents searched for safe escape routes.

The body of Oakland Police Officer John Grubensky was found, along with five civilian fatalities, at a narrow point on Charing Cross Road. It appeared that the cars were jammed at this point by a collision in the narrowest part of the road, and the occupants were unable to escape the advancing flames.

The fatalities included individuals who were unable to evacuate, because of age and disabilities, and several who were overrun by the flames as they tried to escape. Firefighters reported hearing shouts for help from one home and not being able to reach it before it became heavily involved in flames. As their positions were overwhelmed, firefighting crews were split up, and for hours some members did not know the fate of the other members of their companies.

As they pulled out, they tried to evacuate everyone in the path of the fire, and some ended up taking refuge where they could find it. The lieutenant from Engine 19 reported that he was taking refuge with a group at the base of Gwin Tank, using a hoseline to protect themselves as the fire surrounded their position.

A lieutenant and a firefighter had to abandon their patrol vehicle and took refuge in the swimming pool of a hillside home, along with the homeowner, and spent more than an hour under the pool cover, sticking their heads out just often enough to splash water on the cover to prevent its ignition. The house burned, leaving only the pool, and when the fire subsided they found that every house on the block burned to the ground.

Oakland's Division A, BC4, called the incident commander at 1144 hours with the message "fire at both ends – we're going to have to wait it out." The battalion chief had been with a patrol unit that was forced to pull out of Buckingham Boulevard and made its way to Norfolk Road. He left them near the intersection to retrieve his car, directing the patrol on down Norfolk toward Strathmoor Drive to evacuate residents on the opposite side of the hill. The burned bodies of Battalion Chief James Riley and a civilian resident of the area were found hours later, near the location where he was last seen by the patrol unit. It is believed that the chief was trying to assist the woman, who had left her home by car, and both were electrocuted by a falling power line. The message from Division A at 1144 is the last recorded communication from Chief Riley.

Crews on the right flank were also in serious trouble. Engine 8 had been assigned to assist E24, protecting structures in the area of 7140 Buckingham, below the point of origin of the fire, but

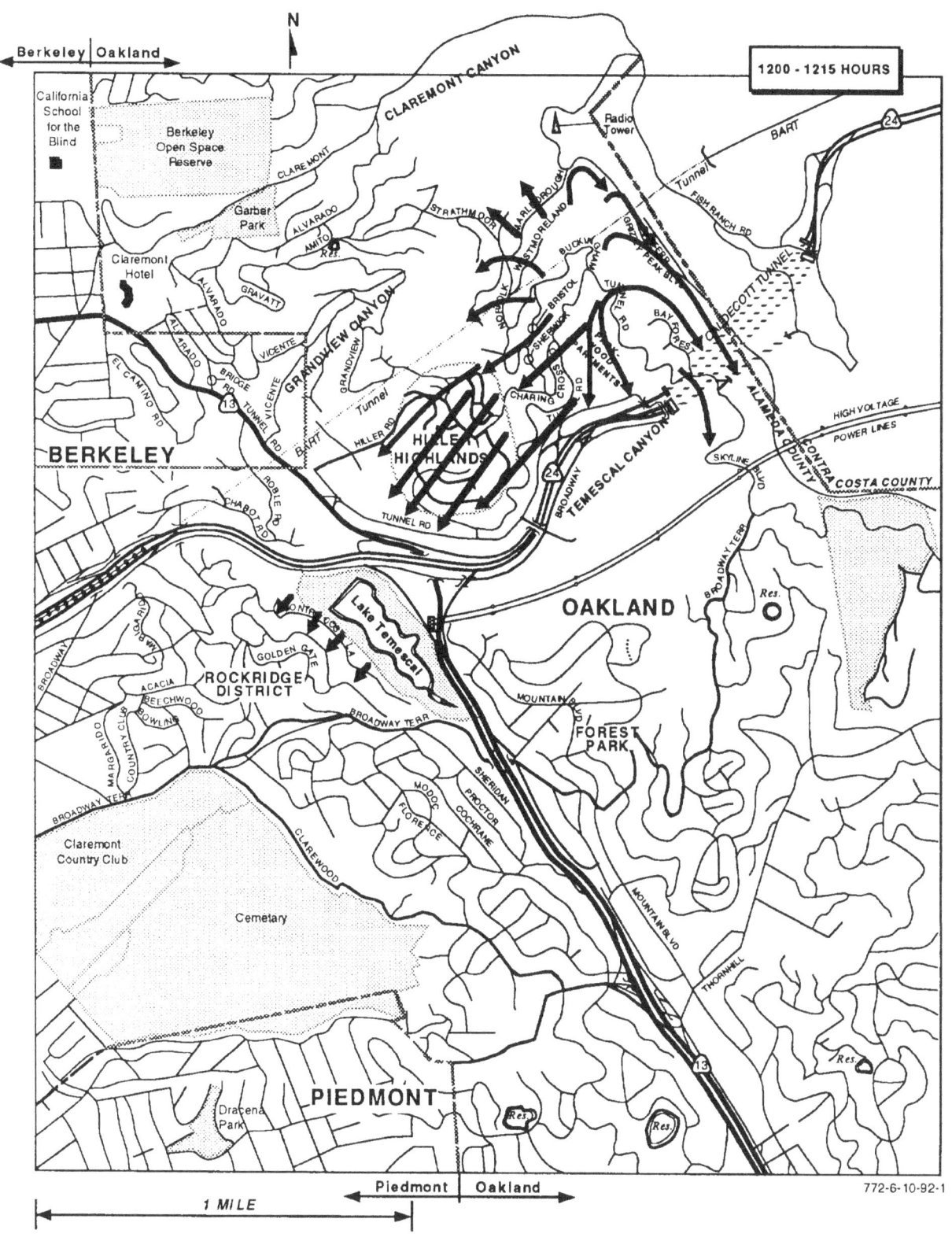

was blocked in the 7200 block by live power lines and abandoned cars. Engine 8's captain directed his crew and a ride-along volunteer to lay a supply line from a hydrant and set up to protect three large houses that were in the path of the fire. They met up with the crew from Engine 24, one crew member from Engine 16, two from Engine 19, and a volunteer who had been helping to fight the fire on the hillside. They were soon joined by five EBRPD firefighters who drove two of their vehicles out of the smoke and flames.

As they set up to protect the structures, the fire that had blocked their path to the west, sweeping down toward the Parkwood Apartments, came back from below their position and cut them off to the east. Seven civilians were trapped along with them. A power line burned through and dropped around the hydrant, burning a hole in the hose and dropping their water pressure. Engine 8 contacted the incident commander to obtain emergency assistance, but there was nothing available to send until the mutual aid strike teams arrived. The captain had his crew members force entry into a large three story house and put all of the civilians inside the concrete block garage. The house was newly constructed and was equipped with an automatic sprinkler system, although he recognized that the sprinklers could never protect the structure from the approaching fire.

The firefighters operated master streams and hoselines on the exterior of the house to keep the wood siding from igniting, flowing an estimated 2,500 gpm, as the fire swept over and around them. Fog nozzles had to be used to protect the members operating the larger handlines that were directed to wet the structure. For an estimated five to eight minutes the fire storm raged around them, burning one of the adjacent houses to the ground, but the exposure protection tactic worked and 20 people survived.

When the storm had passed, they realized that there was still enough fuel left after the fire's pass that it could come back if the wind shifted. They went behind the houses and set a backfire to clear enough area behind to protect them from a rear assault. By this time their hydrant had gone completely dry and they were left with only tank water to maintain their protection.

When conditions finally appeared to stabilize, they met up with the two firefighters and the woman who had survived the fire under the pool cover. Looking down the hill they could see the smoking ruins of the Parkwood Apartments below them; the entire complex had burned to the ground. Then, as they worked their way out of the area, they found the burned bodies of two civilians in the street, only a block from where they had taken refuge. They went on to fight fire for several more hours.

### Situation Status at 1200 Hours

| | |
|---|---|
| *Fire:* | Entire north side of Temescal Canyon involved, up to Charing Cross Road, spreading into Hiller Highlands. Several lives and dozens of structures already lost. East end of Temescal Canyon involved, with fire moving rapidly toward upper Broadway Terrace. Parkwood Apartments fully involved. Fire threatening to top hill near Norfolk Road and spread into Grandview Canyon and to go over Grizzly Peak Drive into Contra Costa County. Flames also threatening to cross Strathmoor Ridge to enter Claremont Canyon. |
| *Resources:* | All companies on left flank in retreat or taking refuge. Too few companies on right flank to make a successful stand. Contra Costa County resources assisting at top of hill, but situation is beyond control. CDF Helitack 106 on scene, making preliminary assessment. Communications systems overwhelmed. Water system failing due to loss of electricity to supply pumps. Additional strike teams and CDF units en route (estimated time of arrival 15 to 45 minutes). |
| *Strategy:* | Firefighters and police officers attempting to evacuate residents and spectators ahead of fire; no resources available to commit to any offensive or defensive action until companies can regroup or mutual aid arrives. |

## All Forces Retreating

When the Orinda battalion chief (BC45) arrived at the command post, he made face to face contact with the incident commander and reported that the Contra Costa County units were working on the upper east side of the fire. The incident commander assigned BC45 to direct the defense of the Parkwood Apartments, but the rear half of the complex was fully involved and companies were still trying to evacuate the last residents from the front buildings. There were no more companies available to assign and nothing could have stopped the fire as it swept through the complex, pushed by the wind coming down the canyon.

The command post had to be relocated as the fire swept toward its position on Highway 24. As the command post was moved to the freeway interchange, a wall of flames was sweeping into the Hiller Highlands development, unchallenged by fire forces. Police officers and firefighters alerted occupants to evacuate moments before a wall of flames enveloped the entire complex.

## Fire Jumps Freeway

At 1202 hours, Oakland Engine 2, which was responding to the main fire, reported to the incident commander that it had discovered a vegetation fire in the Temescal Recreation Area, west of the interchange of Highways 13 and 24. This area is 2,000 feet beyond the face of the burning hills and 400 feet lower in elevation. Flying brands were dropping into this area and within minutes were beginning to ignite a line of trees that border the west side of Lake Temescal. This area was in the direct path of the fire's convective thermal column, which dried and preheated the ignitable fuels.

An off-duty Oakland battalion chief, who had picked up a spare care and was assigned as BC44 to cover the city, saw the fire west of Highway 13 and drove around to the residential area west of the lake. At 1212 hours, BC44 reported to the incident commander that he had multiple structures involved along Contra Costa Lane, a long dead end street that borders the lake. He reported that he was establishing a command post at the intersection of Contra Costa and Buena Vista and initiating operations with Engine 2. Engines 3 and 13 were assigned to stage at Broadway and Golden Gate to assist BC44.

A size-up of the situation west of Lake Temescal was bleak. The fire had jumped the only major physical barrier between the hills and the rest of the city and was now moving into a heavily wooded residential "flatland" area. Flying brands were raining down on the trees and rooftops. The resources of the Oakland Fire Department were fully committed and only a few mutual aid companies had arrived on the west side of the fire. More structures were becoming involved every minute.

Battalion 44 sized-up the situation and suggested calling San Francisco for additional assistance to make a stand. At 1218 the incident commander asked Oakland Communications to request two strike teams from San Francisco. Two minutes later a structure fire was reported on Country Club Drive, several blocks further west, in an area of large homes with a very limited water supply.

The request for two strike teams was made directly from Oakland Fire Communications to San Francisco Fire Communications at 1229 hours. Two strike teams were dispatched and en route to Oakland by 1240.

## Hiller Highlands

At approximately the same time that the fire was jumping across the freeway interchange, it was also sweeping through the 340 unit Hiller Highlands development. There were no fire suppression forces there to make a stand, because all of the companies that had been committed to the left flank of the fire had been overrun or were retreating from the overwhelming fire front. The fire arrived well ahead of the mutual aid strike teams that had been ordered to protect the area. The fire swept over the crest of the hill from Charing Cross Road into the tightly packed two-story town homes, as residents rushed to escape the flames by driving down Hiller Road to Tunnel Road. Several residents of this development were unable to escape ahead of the flames and died in their homes or on the roadways.

## Berkeley Front

On the northwest flank of the fire, flames were spreading quickly toward the city of Berkeley. A portion of the city of Berkeley projects into the lower part of Grandview Canyon, although the major part of the city lies north of Highway 24 and west of the foothills. The city of Berkeley is heavily developed and the part that projects into Grandview Canyon contained closely built two-story dwellings, primarily along Vicente Road and Alvarado Road.

The Berkeley Police and Fire Communications Center received several calls from residents between 1100 and 1130 hours and advised the callers that the fire was in the city of Oakland. (Many of the area residents used Berkeley as a mailing address, although they were actually located within the city of Oakland.) Although no mutual aid request had been received, the Berkeley Fire Department had monitored part of the Oakland radio traffic and was aware of the fire.

Around 1130 hours, the Berkeley duty battalion chief drove to the Vicente Road area and made an assessment of the potential threat to Berkeley. At that time there was no indication of fire moving in that direction.

At 1207 hours, Berkeley received a call from a citizen on Vicente Road, reporting that fire was coming over the hilltop above that address; Berkeley Engine 3 was dispatched to check on that report. Engine 3 drove up Vicente Drive to the intersection of Grandview and Westview before the fire was spotted, coming over the ridge from the area of Norfolk Drive.

Berkeley Engine 3 called for a full first alarm assignment at 1213 hours. Twelve minutes later the companies on Vicente Road were trying to protect structures as flying brands began raining down on them. Dozens of spot fires ignited ahead of the main fire front as it swept down into Grandview Canyon.

| Situation Status at 1230 Hours | |
|---|---|
| *Fire:* | Now out of control on four major fronts. (1) The entire north side of Temescal Canyon, including Hiller Highlands, was already lost, and the fire was spreading into the top of Grandview Canyon and toward the city of Berkeley. (2) Fire was spreading over the ridge into the east fork of Claremont Canyon. (3) The fire on the right flank was still moving to the south, toward the upper reaches of Broadway Terrace and Skyline Drive. (4) West of Lake Temescal, the fire was spreading in the Rockridge district. |
| *Resources:* | Oakland still waiting for mutual aid strike teams to arrive on the west side of the fire. (The first strike team from Contra Costa County was just arriving on the east side and trying to contact the incident commander for instructions.) Two strike teams en route from San Francisco (estimated time of arrival 20 to 30 minutes.) Berkeley Fire Department quickly committing its resources and mutual aid to Grandview Canyon. CDF ground units committing to stop spread into Claremont Canyon from east side. Helicopter 106 working with BC44 in Rockridge district. Two CDF air tankers and spotter aircraft diverted from fire in Contra Costa County (estimated time of arrival 30 minutes). |
| *Strategy:* | Continuing to evacuate ahead of fire. Trying to identify natural positions to make stands ahead of fire when additional resources arrive. |

A Berkeley second alarm was called at 1231, followed by a third alarm at 1238. With all of Berkeley's resources now responding, one engine company had to be diverted to handle a working structure fire in Berkeley. Mutual aid was requested from Albany, Emeryville, and Lawrence Berkeley Laboratory, and off-duty personnel were recalled to place Berkeley's reserve apparatus in-service.

The following map shows the spread of the fire at 1245. The Berkeley units had been driven out of Vicente Road and were setting up to make a stand with master streams in the area of Tunnel Road, Bridge Road, and Alvarado Road. Berkeley established a command post on Tunnel Road and designated the defensive line as Division A. Berkeley companies were also deployed to the Chabot Road area, a quarter mile beyond Tunnel Road where brands had already started an additional flare-up that was spreading from trees to structures. This area was designated as Division B by Berkeley Command.

## CDF Operations

At 1153 hours, Copter 106 was the first CDF unit to reach the scene but could not establish radio contact with the Oakland incident commander. Copter 106 made an aerial size-up of the situation and then dropped its hand crew members to begin manual firefighting on the upper hillside near Grizzly Peak, trying to keep the fire from crossing into Claremont Canyon. The helicopter then began to operate as a water bomber. Shortly after the fire was discovered west of Lake Temescal, Copter 106 redeployed itself to this area, trying to stop the spread on this new front. Lake Temescal was used to refill the helicopter's water bucket.

The first CDF ground units arrived shortly after noon at the Grizzly Peak side of the fire, where they made direct contact with Oakland companies that were engaged along Marlborough Terrace and Grizzly Peak Drive. The CDF battalion chief (BC1616) was unable to establish radio contact with the Oakland command post or with Oakland Fire Communications. Shortly after their arrival, the situation on Marlborough Terrace became untenable and the CDF units backed out to make an assessment of the situation.

Their location put them in the best position to recognize the threat to Claremont Canyon, if the fire continued to spread over the hill and to the north. Unable to make contact with Oakland, the BC1616 decided to commit the CDF units to independent action on the north side of the fire. The worked their way down Claremont Avenue to Alvarado Road and then up to the Amito Road area, where they encountered structures already on fire.

CDF operated independently to cover the Claremont Canyon exposure and established its base at the intersection of Grizzly Peak Boulevard and Fish Ranch Road. Most of the CDF ground resources were committed on this flank of the fire for the remainder of the afternoon. BC1616 coordinated CDF actions with Morgan Hill until a unified command structure was established, around 1600 hours.

CDF was extremely short of resources due to the demands of several simultaneous incidents, including the Franklin Canyon fire that was drawing resources from both CDF and Contra Costa County fire departments. The only two available CDF air tankers and a spotter aircraft had been dispatched to Franklin Canyon at 1156 hours. All other CDF air tankers in the area had been dispatched to other wildland fires in Sonoma County.

The two air tankers that had been dispatched to the Franklin Canyon fire had an estimated time of arrival of 45 to 50 minutes from their bases at Fresno and Salinas. They were diverted to Oakland when it was determined that the East Bay Hills situation was much more critical than the Franklin

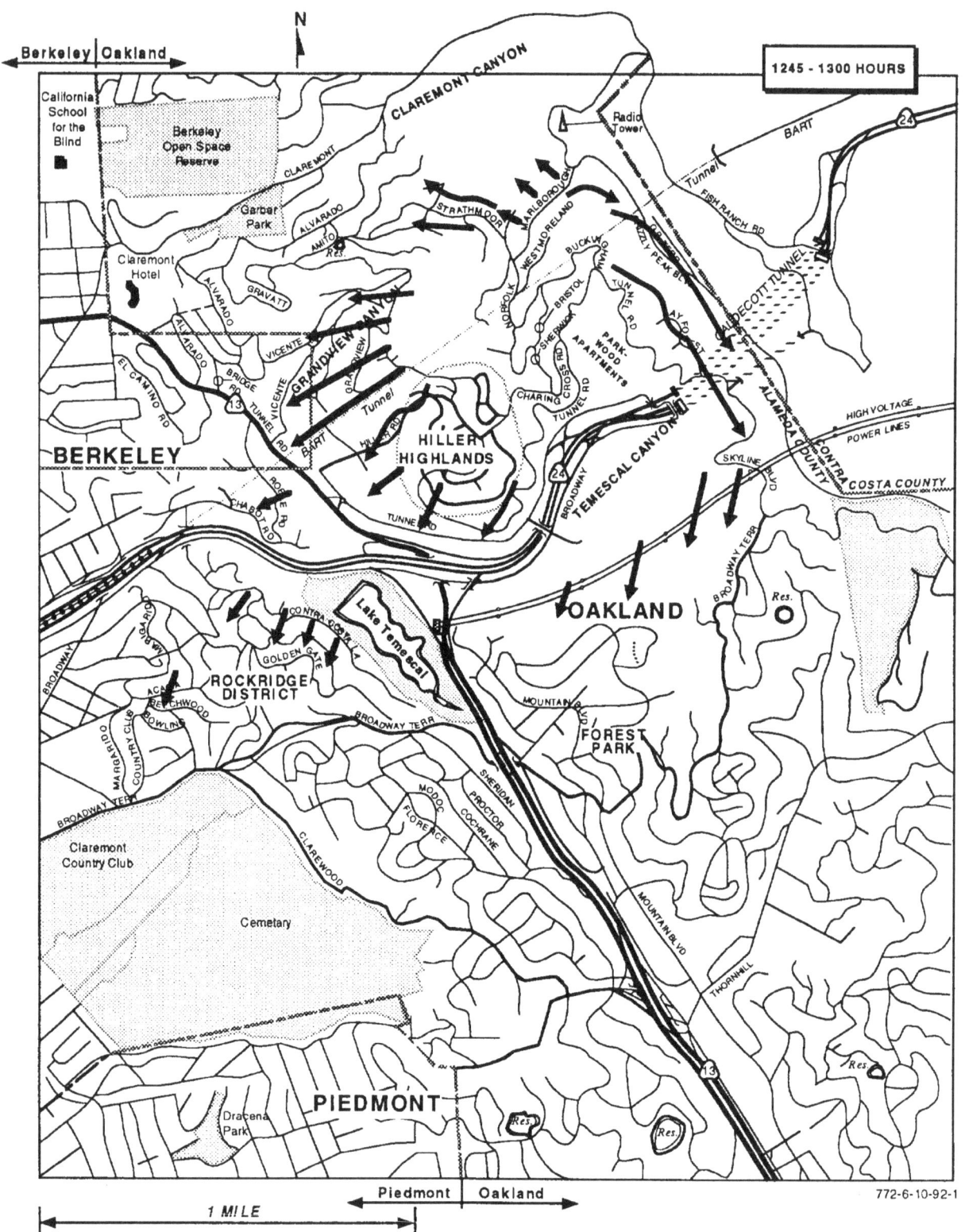

fire. The spotter aircraft was in the area of 1250 hours and contacted air traffic control to clear the air space around the fire for tanker operations.

The two tankers made their first slurry drop at approximately 1300 hours. The tankers tried to drop their slurry on the flanks of the fire, trying to limit the lateral spread, but their operations were restricted by the terrain and the wind, as well as the thermal updrafts caused by the fire. After dropping their loads the tankers had to fly to Santa Rosa, 45 miles northwest of Oakland, to refill their tanks and refuel before returning to the fire.

The need for additional helicopters and air tankers was recognized, and CDF Morgan Hill initiated requests for aircraft from other parts of California. One additional air tanker was dispatched at 1239 hours, and six more were assigned during the afternoon.

**Situation Status at 1300 Hours**

| | |
|---|---|
| *Fire:* | Spreading on five major fronts. (1) Rockridge district, west of Lake Temescal; (2) Grandview Canyon, threatening to jump Tunnel Road; (3) Chabot Road; (4) Claremont Canyon; (5) along the south side of Temescal Canyon, threatening upper Broadway Terrace. Claremont Hotel recognized as a major exposure with possibility of additional spread to flatlands of Oakland and Berkeley. |
| *Resources:* | Alameda county task forces and San Francisco strike teams arriving and in process of deployment; Berkeley Fire Department fully committed to holding action. Oakland and Contra Costa units regrouping on upper hillsides east of fire; CDF ground units deploying along ridge line to protect Claremont Canyon; two CDF air tankers approaching for first air drop. |
| *Strategy:* | Three separate command structures established. Oakland priority on stopping spread in areas (1) and (5); Berkeley in areas (2) and (3); CDF in area (4). All working on establishing defendable lines and evacuating residents ahead of the fire, while hitting spot fires ahead of the main front to prevent leapfrogging. |

## Additional Strike Teams

By 1300 hours, it was evident that additional resources would be needed to stop the progress of the fire in the flatland areas and on the multiple hillside fronts. With the wind coming down from the hills, the fire was being pushed toward the southwest and there were no natural barriers that appeared to be capable of stopping it. Both Oakland and Berkeley recognized the need for more resources at the same time and both requested additional strike teams through Alameda County.

Oakland requested three additional strike teams to respond to Grizzly Peak and one to the intersection of Golden Gate and Acacia in the Rockridge district. Berkeley requested two strike teams to respond to a staging area at Berkeley High School, approximately two miles from the fire. Both requests were logged at the Alameda County Mutual Aid Coordination Center (Lawrence Livermore Laboratory) at 1309 hours. Two additional Alameda County task forces were mobilized, one each from the south and east zones, and the remainder of the request was relayed to OES Region II. Oakland also requested additional command officers to assist in managing the incident, and this request was fulfilled from fire departments in Alameda County.

The strike teams that had been requested earlier were beginning to arrive by 1300 hours. Most had difficulty reaching their assigned destinations and establishing contact with the command structure.

The strike teams that had been directed to stage at Hiller and Tunnel Roads had difficulty with traffic congestion, particularly northbound on Highway 13. By the time they arrived, the fire had destroyed the area they were en route to protect and had jumped over the top of the staging location.

The mutual aid radio channel was hopelessly jammed with communications, and they were unable to make contact with the command post. The command post had been moved to the Highways 24/13 interchange, and some of the strike teams made direct contact at this location.

When the first of the strike teams made their way up Hiller Road, between 1315 and 1330 hours, they found blocks of burning rubble. Every structure in the development was destroyed before any fire suppression crews arrived. The intense fire had already moved on into Grandview Canyon, and the strike teams were able to drive through most of the Hiller Highlands development to Charing Cross Road, where they discovered the bodies of Police Officer Grubensky and several civilians. One badly burned survivor was also located and was transported out to a medical helicopter for transfer to a burn unit.

The companies arriving at the east side of the fire were also unable to establish radio contact with the command post. As they reached the area of Fish Ranch Road and Grizzly Peak Boulevard they found a mixture of Oakland, Contra Costa, and CDF units, attempting to regroup and initiate defensive actions. Most of the resources that arrived in this area deployed to the upper reaches of Broadway Terrace, where they linked up with Oakland companies to protect structures on the right flank of the fire, or with the CDF forces that were trying to stop the fire and protect structures in Claremont Canyon. These battles continued throughout the afternoon as single companies and grouped resources fought a house to house battle with the flames on both fronts.

The city of Piedmont had already committed one engine company to Oakland and had recalled all off-duty personnel to the station. The Piedmont fire chief, while en route to the station, saw fires burning in the Rockridge district, with no fire apparatus or personnel in sight. Recognizing that Piedmont was directly in the path of the advancing fire, he decided to commit the two remaining engines to Rockridge, in hopes of stopping the fire before it reached Piedmont. The Piedmont companies operated in the Florence and Modoc area for the remainder of the day. San Francisco Strike Team 1 was deployed north of the Piedmont units.

One of the San Francisco strike team leaders described the situation on his arrival at the command post, which at that point had relocated to the middle of Highway 24 near the Broadway overpass. "It was eerie – very smoky in the area – almost like night." Structures were burning with a few hundred feet on both sides of the elevated roadway and shrubs in the median were ablaze from a burning brand. As soon as they arrived, the strike team was split to attack the fires on both sides of the roadway, with two engines going to the north side and the strike team leader taking three engines to attack the fire on the south side.

## Claremont Hotel

As the fire continued to spread to the northwest, the huge wood frame structure of the Claremont Hotel became a major concern. The five story hotel, which is believed to be the second largest wood-frame building in the United States, sits on a hillside overlooking the city of Berkeley, at the mouth of Claremont Canyon. The concern was that, if the hotel became involved, it would be a "conflagration breeder," generating a massive additional source of heat and flying brands that could ignite hundreds of new spot fires in the flatland area. Between 1230 and 1300 hours, the fire had swept through Grandview Canyon and was threatening to jump Tunnel Road, a few blocks south of the hotel. Flames were also visible in the upper reaches of Claremont Canyon. A slight shift in the wind would place the hotel directly in the path of the fire.

An Oakland ladder company and a Piedmont engine were the first units assigned to protect the hotel, under the supervision of an Oakland captain designated as Division C. San Francisco Strike Team 2 was assigned to this Division shortly after 1300 hours, and a defensive perimeter was established along the rear of the hotel, utilizing a ladder pipe and several master streams. The lines were supplied from the hotel's private water supply system. The brush and trees along the rear of the hotel were wet down, and the lines were positioned for immediate operation if the fire came down the hill. A second strike team was assigned to Division C to ensure that this perimeter would be held.

With the lines in-place to protect the hotel, the crews began to extend handlines up the hill to Alvarado Road to try to stop the fire on the streets above the hotel.

## New Outbreaks

A set of high voltage electrical lines, supported on steel towers, crosses over the East Bay Hills and drops down to the Pacific Gas and Electric substation that is located in the Y of the freeway interchange, next to Lake Temescal. At 1315 hours, personnel staffing Oakland's command post vehicle, which was parked near the substation, were shocked to see the electrical lines suddenly light up and shower the hillside in front of them with sparks. As circuit breakers popped in the substation, a hasty decision was made to retreat the command post a half mile west on Highway 24 to the Broadway BART Station. This facility, which is located in the median area of the freeway, became the command post for the remainder of the incident.

The arcing of the power lines is believed to have been caused by ionization of the air where the high voltage lines passed through the thermal column created by the fire. The arcing caused the lines to "light up" all the way over the hills. The shower of sparks ignited several new grass fires on the south side of Temescal Canyon, which soon merged with the main body of fire.

The arcing followed the electrical lines all the way to a Pacific Gas and Electric substation in Moraga, and two significant new outbreaks were immediately observed by the CDF spotter aircraft on the east side of the hills. The new fires were within two miles of the main fire and created the risk of two additional fire fronts coming over the hills into Oakland. Four helicopters that had been working on the main fire were immediately diverted to the new outbreaks, with the hope that they could be controlled before they could climb the hills and carry over the ridge to the Oakland side.

ConFire also received numerous reports on the new outbreaks and began dispatching ground resources to the reported locations. One of these fires, known as the Dolores fire, eventually required 49 ground units and was not controlled until the evening hours, after burning 160 acres. Some of the units that were sent to the Dolores fire responded directly from the Franklin Canyon fire as it was brought under control. The other fire, titled the Sunset fire, burned 50 acres and one barn, but was controlled within four hours by nine units.

The new outbreaks delayed the arrival of assistance at the East Bay Hills fire, because Contra Costa resources that were still en route had to be diverted and additional resources had to be requested from the Statewide mutual aid system to cover Contra Costa County. Some of the units worked on two or more of the major incidents in succession.

**Situation Status at 1400 Hours**

| | |
|---|---|
| *Fire:* | Advancing through the Rockridge district and into the Forest Park area. See map on the following page. Heavy action taking place in upper reaches of Broadway Terrace, along Tunnel Road and Chabot Road, and in upper Claremont Canyon. Claremont Hotel still threatened. Risk of fire working out of Claremont Canyon toward University of California Berkeley campus. Two new fires burning on east slope in Contra Costa County. |
| *Resources:* | All available resources in the area already committed. Additional resources hours away. |
| *Strategy:* | Large scale evacuations ahead of fire. Attempting to protect key exposures and hold flanks, looking for natural barriers to make stands. |

## Strategy — Continuing Battle on Multiple Fronts

The battle against the main fire reached a phase that continued for several hours. While aircraft and ground resources were having some success at holding the perimeters of the fire along the north and east sides of the fire area, the flames continued to spread, from house to house and block to block in a southwesterly direction, pushed by the incessant wind. Continuing attempts were made to stop the progress of the fire on three major fronts. In these main directions of fire spread, individual companies and strike teams placed themselves ahead of the fire or on the flanks, trying to stop the flames from advancing to the next house or the next block.

Two of the major actions took place along both sides of Highway 13, in the Rockridge and Forest Park districts. In the Rockridge district, the fire worked its way parallel to the open space provided by the Claremont Country Club and Mountain View Cemetery, heading toward the city of Piedmont. The battle took place south of Broadway Terrace, along Sheridan, Cochrane, Proctor, Florence, and Modoc. A secondary head carried the fire into a hollow where stands were made along Beechwood, Bowling, Margarido, and Country Club Drive, north of Broadway Terrace.

East of Highway 13, the fire worked its way south into a neighborhood at the base of the foothills along Mountain Boulevard. Resources had to be directed south to Thornhill to cross Highway 13, then worked their way back north to make a stand against the advancing flames.

The master stream stand continued along Tunnel, Bridge, and Alvarado Roads, south of the Claremont Hotel, trying to keep the fire from taking the streets at the foot of the hills and to stop it from making an uncontrollable advance into the city of Berkeley and the northeastern corner of Oakland.

Between Tunnel Road and Highway 24, an isolated battle was waged to control the flames that had reached Chabot Road, Roble Road, and El Camino Road. The flames also continued to attack homes on Amito and Gravatt Drives on the upper slopes of Claremont Canyon, above and east of the hotel.

Continuing actions took place on the upper reaches of Broadway Terrace, where the fire was working its way up the steep slopes to attack homes along the southeast perimeter of the fire. In the deeper recesses of Claremont Canyon, the fire was crossing over the ridge through a wooded area, threatening to gain an additional foothold.

## Evacuations

Large areas were evacuated ahead of the fire. Fire officers identified the areas, while the evacuations were conducted primarily by Oakland and Berkeley police officers. The areas that were identified for evacuation were large, because there was no assurance that the fire could be stopped or that the wind would continue to push it in the same direction.

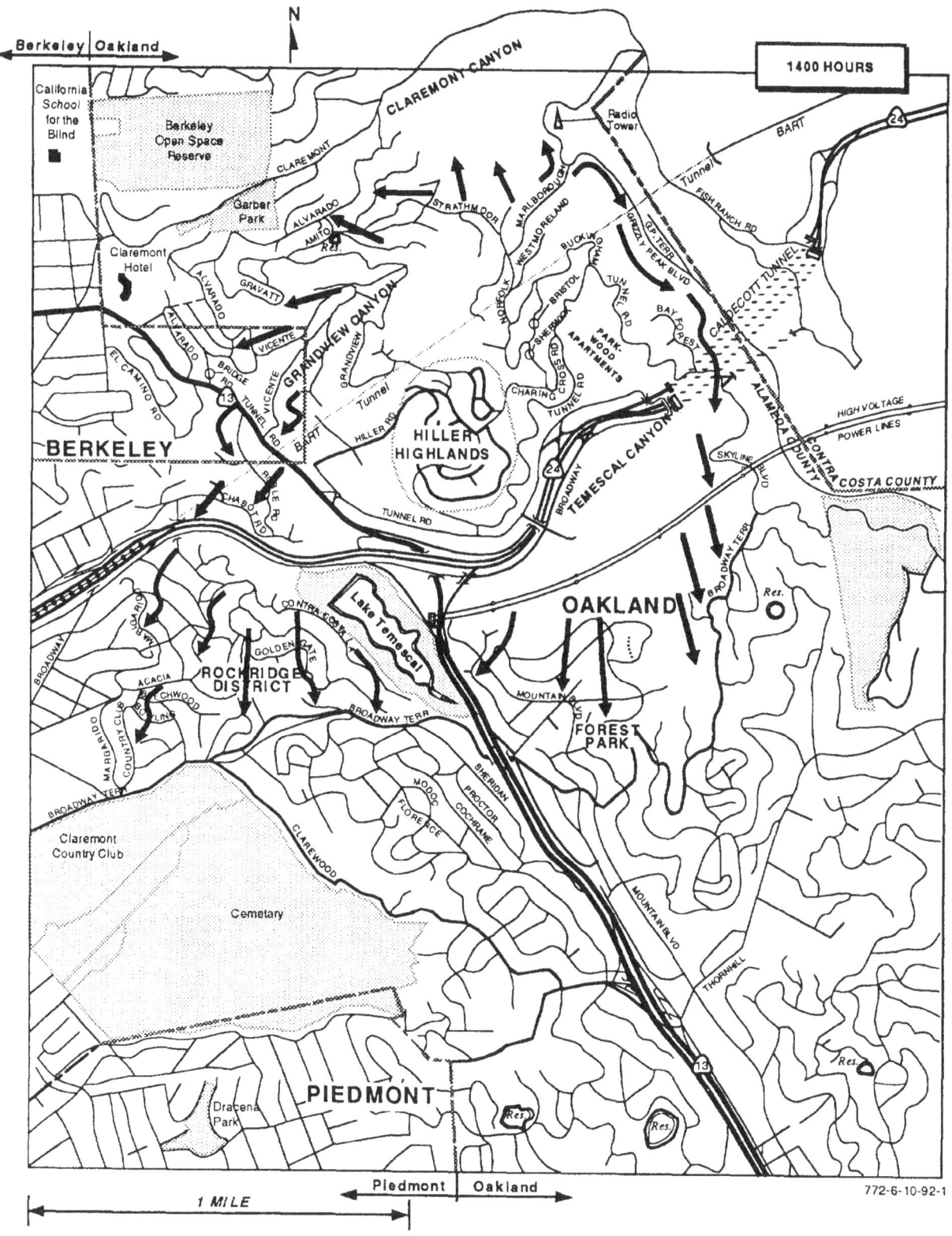

Some of the residents refused to leave and prepared garden hoses and buckets to protect their own homes; others placed garden sprinklers on their rooftops before leaving. A command officer, who had been assigned to a helicopter, described the reactions of residents who were advised to evacuate over the helicopter's public address speakers. While some hurried to leave the area, others displayed their determination to stay and their contempt that the fire forces were not doing enough to protect their properties. Police officers had to use their authority to evacuate some residents and to keep others from returning to endangered areas.

The evacuation included more than 5,000 people from significant portions of Oakland, Berkeley, and Piedmont, as well as part of the University of California Berkeley campus. The cities' emergency planning and support agencies were activated to open shelters and provide food and other services. The basic logistics of evacuating a large urban area are complicated and required a large commitment of personnel, primarily from law enforcement agencies.

## Tactics

Two basic tactical approaches were used. The structurally-oriented strike teams used 1-1/2-, 1-3/4- and 2-1/2-inch handlines, as well as master stream appliances where sufficient water was available from hydrants. The wildland-oriented strike teams used 1-inch attack lines, supplied by tank water, and were much more mobile. While the mobility was important in controlling the numerous spot fires that were ignited by brands and embers ahead of the main fire front, the heavy streams were needed to have any effect where the main body of the fire was creating the exposure. Both tactical approaches were effective in different situations, but neither one could stop the progress of the fire in the downwind direction.

A few homes were saved by homeowners who made determined efforts to protect their properties. On Margarido Drive, a retired battalion chief enlisted his off-duty firefighter son-in-law to keep the flames from igniting his home. In another area, three homeowners banded together to protect their homes as they were exposed in succession. Most of the structures that survived were the products of determined efforts by firefighters who identified locations they could defend and stayed until the threat had passed, before moving on to another location. The few homes that were left standing demonstrated the value of noncombustible roofing materials and brush clearance around the structures.

Company officers reported that their efforts were most successful when they could group several companies together and make a coordinated effort to save a group of homes. They also reported the frustration of fighting to save a home for an hour or more, only to see it burn when they ran out of water or were simply defeated by the fire's unyielding energy.

Firefighters on the front lines endured hours of frustration and punishment. The 90 degree weather and the hot dry wind were enough to cause heat exhaustion at a routine fire; this battle continued for hours without relief, in the face of a fire that drove heavy smoke, flaming brands, and embers into the faces of the fire suppression forces. Handlines had to be moved repeatedly to hit new outbreaks on rooftops and in vegetation. Interior attacks were attempted to hit wood shingle roof fires from below. Houses burst into flames and fire swept across streets all around them. Time after time, the firefighters fought until they won or had to retreat, then moved on to the next block and did it all over again.

One of the Alameda County Task Force commanders reported that his convoy was en route to an assigned staging location at Golden Gate and Acacia in the Rockridge district, but had to take an indirect route due to blocked roads and traffic congestion.  At Proctor and Broadway Terrace they were flagged down by a resident urgently requesting assistance to evacuate a disabled relative from a nearby home.  Before the rescue could be completed, they found themselves fighting flaming brands and embers that rained down on the trees and rooftops around them, igniting dozens of fires within the block.  Unable to contact the command post, they fought to keep the flames from involving one house after another until they had to pull back and retreat to the next block.  The rescue was completed, but the block was lost.

As the battle continued, they developed tactics that were successful in protecting individual homes, where they could deploy ahead of the fire, but the majority of the homes burned to the ground around them.  Time after time they pulled back to make another stand, first on Proctor, then on Agnes, Florence, and Modoc.  They saved one or two homes where they could, but they were never able to stop the fire's advance.  It was hours later before they discovered that their successive positions had been directly on the head of the advancing fire.

## Assessment of the Situation

The critical shortage of resources was the major problem for several hours.  As soon as any single or grouped resources would arrive, there were assignments waiting for them.  Command post personnel were trying to obtain good information on the fire's perimeter, trying to predict the direction the fire would take in order to identify evacuation areas, and prioritizing requests for assistance that came from several different areas in rapid succession.

It was difficult to determine resource deployment because so many units were engaged in actions that were unknown to the command post.  They were engaged in different areas, had no radio contact with the command structure, and were operating on their own initiative.

Efforts were being made to develop an incident command structure and to build a logistics system to support expanding operations.  Division assignments were being made to supervise operations in particular areas that could be identified, but there was no information on where many companies were working, what conditions they were encountering, and what success they were achieving.  One Oakland captain reported that when he finally reached the command post, after the situation calmed down in his area, he found out that he had been assigned, hours earlier, as a division supervisor over several other units.

The battalion chief who had been assigned to make an aerial assessment from the police helicopter had to return to the command post to deliver his report.  This information led to the conclusion that massive additional mutual aid resources would be needed to contain or control the fire.  Alameda County resources had already been severely depleted and Contra Costa County was requesting assistance from other areas to respond to the fires that had broken out on the opposite side of the hills.  San Francisco was requested to send additional strike teams, but there was a concern that flying brands would begin to ignite fires in that city on the opposite side of the bay; the smoke was already banking down and causing problems in San Francisco.  After sending 25 percent of its on-duty forces to Oakland, San Francisco was calling back off-duty personnel to increase staffing on its remaining companies.  San Francisco was able to send several special units and additional personnel to Oakland.

With local resources already depleted, the assistance would have to come from distant locations with extended travel times.  The overall incident strategy was changed to a "campaign approach," based on assembling the resources that would be needed to deal with a worst case scenario.

At 1359 hours, Oakland requested 13 additional strike teams, 6 air tankers, and 6 helitack units. The ground units were directed to a staging area at Raimondi Park, three miles from the fire, near the Oakland end of the Bay Bridge. This location provided the space for a full base camp operation to support the incident and a large open area to service helicopters. Thirty minutes later Berkeley requested two more strike teams to respond to the Berkeley High School staging area. This large scale assistance would take from one to four hours to arrive at the staging areas.

The fire was headed straight for the city of Piedmont, which had already committed all of its apparatus and personnel to Oakland. All that was left in Piedmont was a Streets Department street flusher, staffed by one firefighter. Piedmont requested the assignment of strike teams to protect the city, but none could be allocated until the distant mutual aid forces could arrive and the most urgent requests could be accommodated. (A strike team was briefly assigned to cover Piedmont, later in the afternoon, but it too was soon committed to fighting fires in the Rockridge district. The Piedmont companies returned to their city around 0100 hours and stood-by at the perimeter for the remainder of the night.)

## Unified Command Structure

Around 1600 hours, implementation of the incident command system (ICS) on a large scale began to bring the incident into focus at the command post. The map on the following page shows the extent of the fire by this time. Prior to this point, Oakland and Berkeley had each operated independently, with their own command posts. A Berkeley officer had been assigned as a liaison at the Oakland command post. The CDF units were working under their own command structure, primarily on the north and east sides of the fire, with limited contact between CDF and the Oakland or Berkeley incident commanders. San Francisco had established a secondary command post at the Claremont Hotel to manage its resources, which were operating entirely on their own radio channels, with a liaison officer assigned to maintain contact at the Oakland command post.

A unified command structure was implemented, involving Oakland, Berkeley, and CDF command officers, and the overall incident was restructured to divide geographic and functional responsibilities into manageable components. It took several hours to make a complete assessment of the situation and to develop an organization structure to deal with the complex incident. To simply cover the perimeter of the fire, three branches and 15 divisions were established.

Branch 1 incorporated the area from the Claremont Hotel to the top of Grizzly Peak.

Branch 2 covered the area along the south side of the fire, from Grizzly Peak to Highway 13.

Branch 3 covered the Rockridge district, from Highway 13 around to the Berkeley front on Tunnel Road.

To assist in establishing an effective incident management structure, two caches of 40 portable radios each were requested from the Alameda County Mutual Aid system. These radios provided additional channels to organize the communications network for a large scale incident.

An "Overhead Team" was also requested from the State Office of Emergency Services (OES) to bring in personnel experienced in managing large scale incidents, including specialists in logistics, communications, and finance, along with supporting equipment. The Overhead Team would take several hours to arrive.

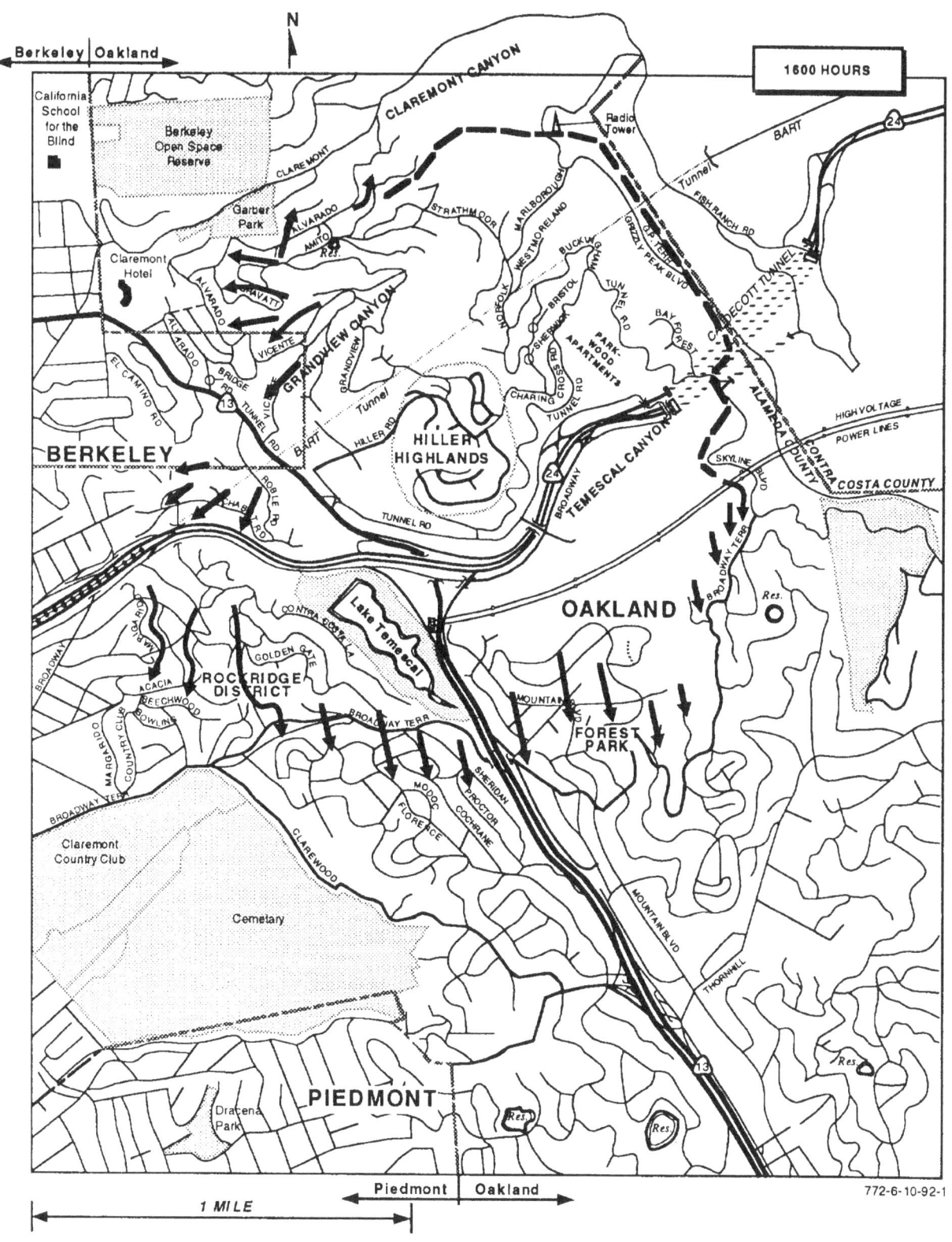

## Additional Resources Ordered

After a further assessment of the situation, 11 additional strike teams were requested at 1746 hours, also directed to report to the Raimondi Park staging area. To obtain this level of resources, fire departments from up to 300 miles away mobilized units to respond to Oakland. These forces would be needed to relieve exhausted firefighters, some of whom had been engaged on the front lines for almost seven hours. The extent of the fire at approximately 1800 is shown on the following page.

The San Francisco strike teams were able to maintain contact with their own communications center and requested additional support from their own department. Two hose tenders were activated by reassigning ladder company crews and responded to the staging area at the Claremont Hotel. These units are loaded with 5-inch hose and are equipped to establish emergency above-ground water supply systems in the event of an earthquake. Three water tenders and a communications vehicle from the Department of Public Works were also dispatched to Oakland. In response to further requests for assistance, San Francisco called in 85 additional off-duty personnel and sent them to Oakland by bus to reinforce the companies at the Claremont Hotel and in the Rockridge district.

On the hill above the Claremont Hotel, the fire was advancing through a mixture of modern homes and stately mansions, working its way down the hillside. With the hotel protected by master streams, poised for action, the crews climbed the hill to engage the fire on the upper streets, but found that the hydrants on the upper streets were dry. They returned to the bottom and started up again, hand-stretching a 5-inch hoseline to support an offensive attack on the fire. The fire was successfully held at Alvarado Road.

In the Rockridge district, the 5-inch hose and portable hydrant system was used to bring a strong water supply into the area where San Francisco units were operating. Several companies were able to obtain water from the portable hydrants.

## Wind Changes

The main battle continued until approximately 1930 hours, when the wind finally abated. The smoke and heat changed from pushing ahead of the fire to rising vertically as the wind eased off. Within a few minutes, a cool damp ocean breeze began to push the products of combustion back into the burn area. This stopped the uncontrollable advance of the fire, but left a huge perimeter of blazing homes that continued to expose adjacent structures.

The battle against these fires continued well into the night around the entire five-mile perimeter of the fire. From time to time the wind would pick up, and the burning intensity would increase for a few minutes, showering the area with new sparks. Crews worked to build a safe perimeter around as much of the fire as possible, including attacking structure fires and shutting off burning gas lines.

CDF hand crews were used in some of the wooded areas, particularly in Claremont Canyon and along the Piedmont border, where fire had spread to a wooded hollow next to the cemetery. Bulldozers were used to clear a fire break in Claremont Canyon.

The major concern was that the wind would return in the morning, so plans were made to bring in even more resources, to overhaul as much of the perimeter as possible and to be prepared for a new outbreak. A priority was placed on relieving units that had been in operation for as long as 10 hours. The strike teams that were arriving from the earlier requests were assigned to relieve tired crews. A request for 30 additional strike teams was made at 2030 hours, plus 10 more helitack units

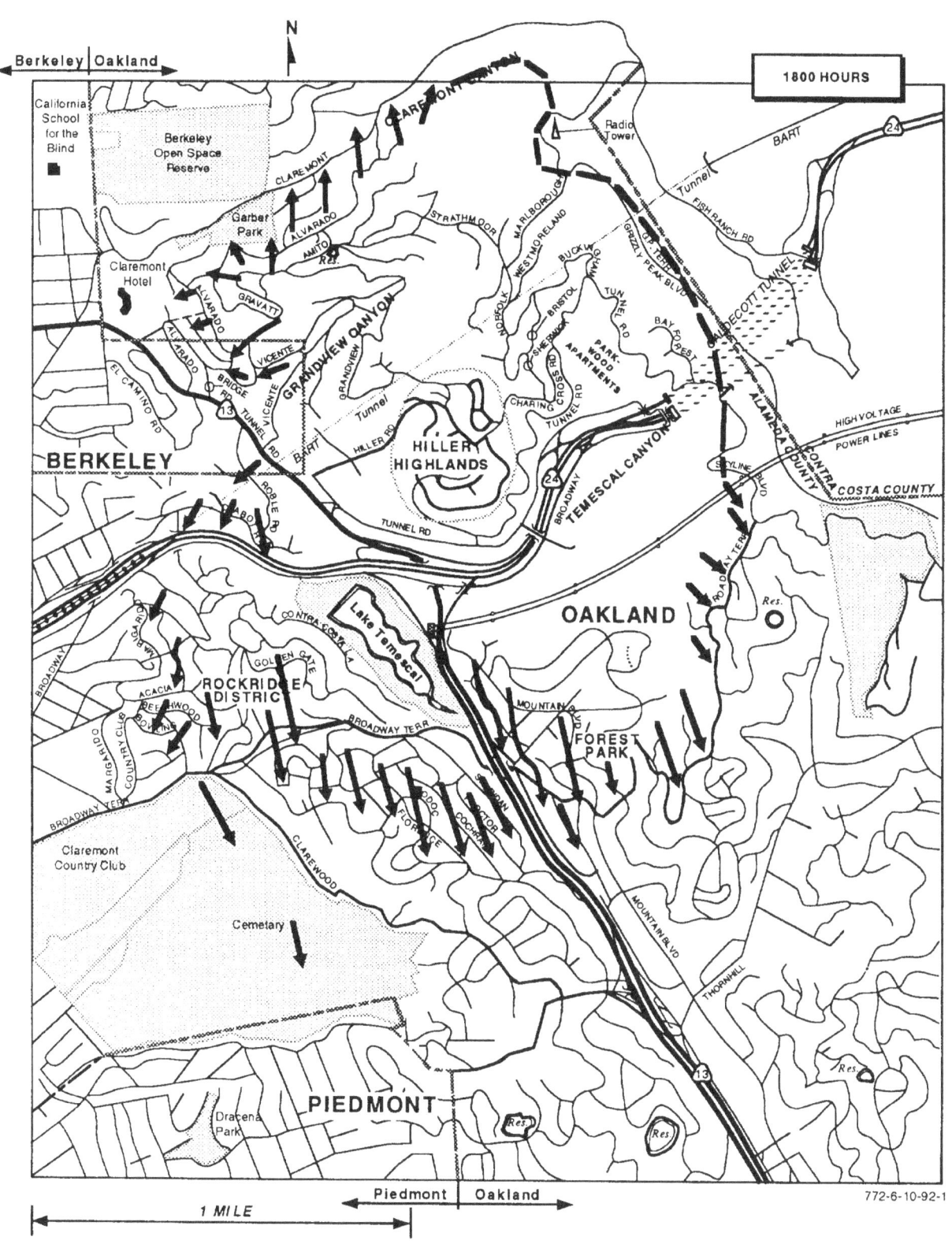

at 2200 hours. Berkeley requested two additional strike teams to report to its staging area. These resources were ordered with the anticipation that they would travel during the night and be available for assignment in the early morning hours.

The OES Overhead Team arrived at the command post during the evening to assist with the development of a structure for extended management of the incident, mobilizing State government resources to support Oakland. The staging area was moved to the Alameda Naval Air Station around 2200 hours, where facilities were available to support a base camp for several hundred firefighters and an inactive runway area could be used to stage apparatus.

## MONDAY THROUGH THURSDAY

During Sunday night, a helicopter with infrared monitoring equipment was used to survey the fire area, and the data was transferred to a geographic database map. This information was used to plan the deployment of resources for Monday morning and to identify the most vulnerable areas if the strong winds returned. There were still dozens of structure fires burning in different parts of the fire area and flames were evident in some wooded areas. Thousands of smoking ruins marked the locations of obliterated homes. As long as these fires persisted, the fire continued to be a threat. The first objective was to contain the fires within a safe perimeter, then to extinguish all the remaining flames.

Daylight offered the first opportunity to survey the fire area from the air under reasonably calm conditions, although the atmosphere was still heavily laden with smoke. Fortunately, the wind did not return on Monday morning, and weather conditions were favorable for developing a secure perimeter and eliminating potential sources for rekindles. Water supply was restored to some areas when East Bay Municipal Utilities District personnel brought in emergency generators and transportable pumps. Public works and utility crews had to clear the streets with heavy equipment to provide access to several areas.

The fresh strike teams were deployed and most of the units that responded during the first six hours were released. This restored normal protection to most of the communities in Alameda and Contra Costa Counties. Units from all over California took part in the continuing operations at East Bay Hills.

Attempts were made to locate bodies and account for missing persons in the burned area, and damage assessment and preliminary investigative efforts were initiated. More than 100 people were initially reported as missing, mostly by friends and relatives who were unable to make contact with people who lived in the fire area. Many families were separated during the evacuation and were unsure if everyone had reached safety. Lists were compiled and search efforts were directed to the areas where missing persons were last seen.

### Losses

The fire was declared as contained on Tuesday morning, with overhaul continuing throughout the day. The final, full extent of the fire is shown on the following page followed by an aerial photograph taken after the fire.

Reports of missing persons were processed, attempting to identify bodies that had been located and gradually searching areas as overhaul continued in the burned areas. Most of the bodies that were recovered from the rubble were so badly burned that they were difficult to identify as human remains.

Within an area of more than two square miles, only a few structures remained standing. Even people familiar with the area had difficulty navigating, as all landmarks were gone and the appearance of the entire area was radically changed. Crews worked their way in from the perimeters, completing a thorough overhaul of the rubble. Where personal belongings could be salvaged from the rubble, they were turned over to the owners, but most of the homes were so completely destroyed that few items could be salvaged.

As the extent of the destruction became evident, the area was declared a Federal Disaster Area. The Presidential declaration made Federal relief funds available to reimburse agencies for their costs and to help in rebuilding.

Most of the strike teams from distant areas were released, primarily on Wednesday and Thursday. Ten additional OES engines were brought in for continuing overhaul, along with two urban search and rescue teams with dogs to continue the search for bodies in the rubble. The final death toll was set at 25, with approximately 150 injuries to civilians and firefighters.

After the full area was surveyed for damage assessment, the number of people left homeless was estimated at close to 10,000 and damage estimates exceeded 1.5 billion dollars. The actual number of structures destroyed was eventually determined to be 3,354 single family dwellings and 456 apartment units. Approximately 2,000 burned vehicles were also located in the area.

## ANALYSIS

### Fire Risk

The most significant factor that should be recognized from this incident is that the fire was beyond the capability of fire suppression forces to control. The stage was set by a number of contributing factors that created the opportunity for disaster. When the Santa Ana wind condition was added to those risk factors, the combination was more than any fire department could handle. It was remarked by one fire official that if the same fire risk factors had been present in a park or forest, the area would have been closed to all activities. As long as the wind was present, the fire was going to continue to spread, no matter what strategy and tactics were used and no matter how much equipment and how many firefighters were there to try to stop it. The fire was contained only when the wind changed.

The spread of the fire during the first hour is virtually unprecedented for an urban conflagration, including wildland-urban interface fires. Just 64 minutes after the fire broke out, it was burning in the Rockridge district, more than a mile downwind from the point of origin. This rate of fire spread and the difficult access to the fire area meant that firefighters arriving to combat the fire could not be effective against it, and they were in extreme danger from it.

The factors that created the extreme fire risk situation in the East Bay Hills are not unique. There are similar situations in many parts of the United States and particularly in the coastal areas of California, which have seen some of the most destructive wildland-urban interface fires over the last 70 years. This incident may be viewed as "the ultimate" interface fire, but there are many reasons to believe that it could easily be repeated or surpassed, unless a major hazard mitigation effort is instituted.

The risks can be significantly reduced in several ways, all of which have been identified and advocated for decades by the fire protection community. There are political and economic reasons why these recommendations have not been implemented. After the fire there were two opposing forces

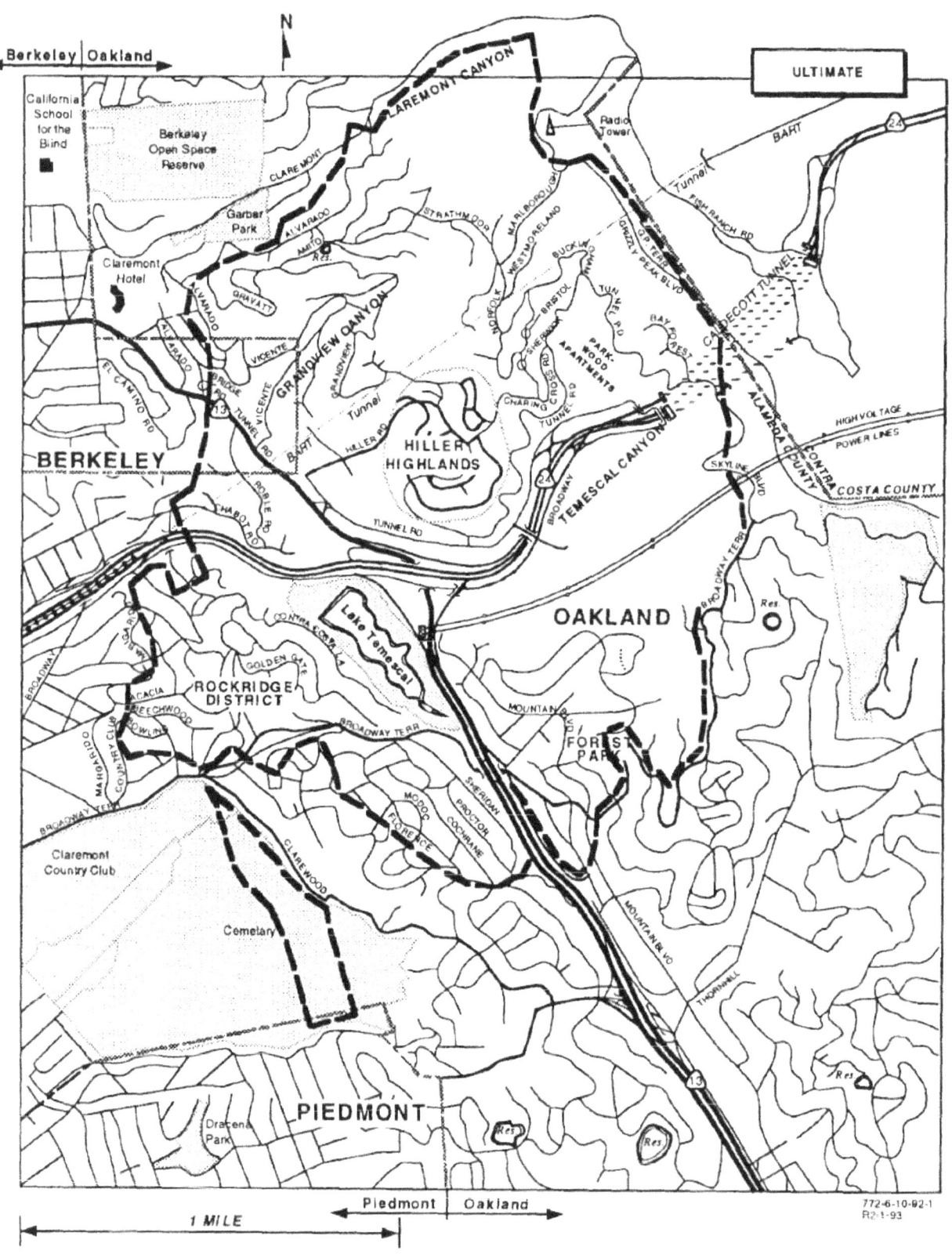

Aerial photograph taken after the fire. Area shown is the same as on page 9.
In total, 3,354 structures were destroyed.

at work in Oakland; those who wanted the hill area rebuilt and restored to its former condition without delay and those who wanted to wait until the risk factors could be mitigated before allowing any rebuilding. There was also a deluge of litigation relating to lost lives and property, claiming that various public and private entities were negligent in their responsibility to prevent or control the destruction.

Without risk mitigation efforts, the same area could easily become vulnerable to another disastrous fire. The adjacent areas, which were spared in this event, are equally vulnerable, as are dozens of areas in California and other parts of the country with similar characteristics.

The lesson from this fire should not be that public fire protection forces were unable to control it. The more important lesson is that the risks were recognized and the consequences were accurately predicted long before the fire, but nothing was done to mitigate the risks before the fire occurred.

## Fire Origin

The Saturday afternoon fire was controlled by aggressive fire suppression efforts, involving a major commitment from the Oakland Fire Department, assisted by EBRPD, CDF, and Berkeley units. It was a "good stop" in a very difficult location. Crews continued to overhaul the burn area until darkness made it too dangerous to operate on the steep hills and the scene was checked during the night for signs of hot spots or rekindles, which would have been readily apparent in the dark. Even knowing the tendency of fires to burn down into duff and root systems, there is little more that could have been done before the following morning to continue the overhaul process.

On Sunday morning, the wind made the hotspots evident. Units were sent to complete the overhaul and experienced officers were confident that the situation was under control. The sudden eruption occurred when a firefighter was digging out a hotspot, near the perimeter, and sparks were carried into an area of dry brush, which virtually exploded into flames. The sudden development of the new fire was witnessed by several individuals, including experienced firefighters, who were able to describe the phenomenal rate of fire growth and spread.

## Fire Characteristics

In this case it is evident that the wind played a major part in every aspect of the fire's growth and spread, when the wind died down, the fire's progress was stopped.

The East Bay Hills fire has been described as a conflagration and as a fire storm in media accounts. The synergistic efforts of the wind and the thermal energy released by the fire created unusual fire phenomena that exhibited some of the characteristics of a fire storm on a localized basis, but the term conflagration is more appropriate for the overall situation.

A fire storm is defined as a situation in which the fire's thermal energy creates its own weather phenomena, overpowering natural meteorological forces. There were many observations of flaming whirlwinds, crowning fire in the treetops, and rolling clouds of fire moving through the air or along the ground. Firefighters described balls of fire floating through the air around them. These descriptions are evidence of localized fire storm characteristics. One witness described the fire in Hiller Highlands as a single swirling mass of flame, involving buildings, trees, and vehicles, consuming everything as if it was in a gas oven. This area was totally consumed by the fire, which spread so rapidly that residents died in the streets trying to outrun it.

Natural convective forces cause a fire's plume of smoke and superheated gases to rise. A fire of this magnitude releases tremendous quantities of thermal energy into the plume every second. It is estimated that a wind between 15 and 30 miles per hour is sufficient to bend the thermal column and cause it to travel parallel to the ground, projecting out ahead of the fire. The superheated gases preheat exposed fuels, hundreds or possibly thousands of feet ahead of the fire, causing them to burst into flames with almost explosive force when they are ignited. The ignition may come from a flying brand or a glowing ember, or the preheating may continue until materials reach their auto-ignition temperatures. The ignition process is similar to the effect of a thermal layer of superheated gases within a room that radiates heat down onto the exposed contents prior to flashover.

One phenomenon that was observed at this fire was the ignition of the tops of wooden power poles ahead of the fire. The tops of the poles were high enough to project into the thermal layer and were ignited by convective heat transfer over the heads of firefighters working below. This suggests that the firefighters were working in an area that was being preheated by radiant heat transfer from the superheated gases above, as well as from the approaching flame front.

The actual spread of the fire, in most cases, was observed to be flaming brands and embers, carried by the wind and dropping onto ignitable fuels ahead of the fire front. The ignitable fuels included trees, brush, grass, and other natural fuels, as well as wood roofs, debris in rain gutters, and other combustibles around structures. The preheating process dried any remaining moisture from the fuels and may have elevated their temperatures close to their auto-ignition point before the brands or embers landed. When an open flame came in contact with these preconditioned fuels, they would become involved in a fraction of the normal time.

In many cases the embers, blown by the strong winds, were seen to work their way in under the eaves of houses, readily igniting even tile-roofed structures. Where the walls and roofs resisted ignition, the exposure caused by burning trees and brush was often sufficient to cause interior ignitions by radiant exposure through windows. The worst case were the hillside homes, where natural fuels carried the flames directly under overhanging structures.

Very few structures in the area escaped total destruction. Most of the houses burned to the ground, as the fire totally consumed all available fuels. The combustion process was very rapid, as it took place in a superheated environment with a constant supply of fresh air, so houses became fully involved very rapidly and were totally consumed in less than an hour. The same observation was made of the more than 2,000 vehicles that were burned in the fire area; in most cases only a hulk of corroded steel and melted glass was left behind.

The most rapid fire spread occurred during the first hour of the fire. Temescal Canyon provided a natural draw for the fire with the wind blowing in over the ridge and down into the canyon, then sweeping out toward the open end. This action split the fire into two fronts and then carried the fire along the north slope, spreading more than a mile in the first hour. This initial spread carried the fire all the way to Hiller Highlands, before it jumped 2,000 feet of freeway interchange and open space to begin its run on the other side of Lake Temescal. In comparison, the head of the fire advanced only three quarters of a mile in the next seven hours in the densely built-up Rockridge and Forest Park neighborhoods.

## Wildland-Urban Interface Characteristics

The East Bay Hills could be described as an extreme example of a wildland-urban interface zone, where the fuel supply was an intimate mixture of natural trees, brush, and grass surrounding artificial structures and vehicles. The complete intermingling of the natural and artificial fuels, combined with the steep terrain, created a combination that was more hazardous than either urban or wildland fuels alone. Hundreds of homes were completely enveloped in an extremely flammable environment. The natural fuels provided a continuous fuel blanket to carry the fire across the hillsides.

The fire differs from many previous interface fires in that it originated within a developed area. Most of the major interface fires have originated in more remote wildland areas and grown to major proportions before attacking urbanized areas. This fire originated within a few hundred feet of occupied homes and involved structures in the first 10 to 15 minutes, in spite of the fact that firefighters were present when it broke out.

There was no time or space to attack the fire before it involved the urbanized area and no time to establish a defensive barrier ahead of the fire. The combination of wind and thermal forces turned water streams away when they were directed on the fire, forcing the firefighters to abandon any early attack strategies and concentrate on evacuating residents and themselves from the path of the flames. The fire moved so quickly and grew so fast that firefighters were in imminent danger trying to evacuate the residents ahead of the flames.

### Initial Response

Initial response time to this fire was not a factor, because firefighters were actually present and working on hot spots from the previous day's fire when this fire broke out. Response time to this area of the hills is normally a concern, because fire apparatus must climb the steep hills on narrow switchback roads. The access problems caused a delay for companies responding to reinforce the units that were already on the scene, but there is no evidence to suggest that this was a significant factor in the outcome of the fire.

Analysis of the fire strongly suggests that it was uncontrollable by conventional firefighting methods within the first few minutes. The fire broke out and increased in size so rapidly that the firefighters who were already present with charged hoselines could not stop it and had to retreat for their own safety. With the Santa Ana wind blowing into a canyon that was so ripe for fire propagation, the fire would have overwhelmed any conventional fire suppression capability. This had been recognized and predicted years earlier and was commented upon by the officers who had directed the previous day's incident, when the wind had been the only missing factor.

## Training And Preparation

The Oakland Fire Department is and has always been a structure-oriented fire department. Before budget reductions of the 1970s and '80s, it was recognized as one of the strongest fire suppression departments in the western United States. The budge limitations reduced the number of companies in service and the staffing on each company. Several stations were closed during this period.

Wildland fire suppression was not considered to be a major area of emphasis for the department, although many of its officers have developed knowledge and expertise in this type of firefighting. The department has not been directly involved in some of the programs that have been developed for wildland fire suppression forces, and this caused some problems with the integration of operations at the fire.

Oakland, as the largest city in the area west of San Francisco Bay, is seldom in the position of requesting mutual aid. There were problems with the processing of mutual aid request, partly due to problems with terminology and procedures in the communications center. The shortcomings of the communications system were also a major obstacle to effective incident management. The radio system proved to be inadequate for the scale of operations that was necessary, even for the initial stages of the incident. These factors made effective coordination or control of the mutual aid resources that arrived during the first five hours impossible.

## Incident Management

The management of an incident of this size and complexity is a tremendous challenge. From the beginning of the incident, the situation expanded and changed more rapidly than the suppression forces could communicate, obtain reinforcements, and get organized. Almost 800 structures were ignited within the first hour and more than 300 per hour for the next seven hours. These factors created a situation that exceeds all previous experience with ICS or any other incident management system.

After the Loma Prieta earthquake in 1989, the Oakland Fire Department placed an emphasis on the full implementation of the ICS system for managing incidents and coordinating mutual aid. The ICS system was used for this fire, but the shortage of command officers and the extremely fast escalation of the incident made it very difficult to develop the organization in proportion to the situation. With the radio system overwhelmed and companies in retreat or taking refuge, the early attempts to organize the operation were unsuccessful. There were not enough command officers to assign manageable areas, size up the situation, develop strategy, direct tactics, or to even account for the location and actions of the resources that were deployed.

The officer in charge of Engine 19 was in command and had a good view of the situation when the fire erupted. He initiated calls for multiple alarms and mutual aid, provided a good size-up report for the responding assistant chief, and directed the first responding units to critical locations.

The assistant chief was extremely familiar with the locations and the risk factors and was on the scene within the first 12 to 14 minutes, but the fire was already so big and spreading so that there was no opportunity to make an accurate assessment of its size, direction of travel, or rate of spread. The position that was the initial choice for a command post location turned out to be in the direct path the fire took and was too dangerous to use. The fire was obviously moving quickly toward occupied dwellings and growing at an almost unbelievable rate. The severity of the problem and the extreme danger were recognized immediately, but with the fire spreading in three directions at the same time, there were insufficient resources available to cover the critical exposures. There was not even a safe vantage point to size up the situation.

The first arriving battalion chief was assigned to supervise the area that appeared to be the most critical exposure, while the assistant chief tried to make a size-up by moving around to different vantage points. The situation required several more command officers in those first few minutes to establish divisions in the most critical areas in order to make an assessment of conditions and supervise the actions of units. Based on their information, the incident commander could have determined the best strategy to employ and prioritize the assignment of resources. The shortage of command officers made it impossible to perform these standard command functions. The lack of aides to work with the command officers compounded the problem, because there was no one to assist with recording information, managing radio traffic, or utilizing alternate radio channels and communications systems.

The shortage of command officers also compounded the communications bottleneck. The incident commander had to deal with information coming from too many different sources and had to direct too many individual units. The capacity of the primary assigned tactical radio channel and the mutual aid channel were soon overwhelmed and the communications process broke down.

When an off-duty assistant chief arrived, he was in a better position to assume overall command of the incident and establish a command post at the bottom of the hill, while the on-duty assistant chief attempted to direct the Operations section from the top of the hill. Smoke conditions made it impossible for the incident commander to see where the fire was going from the command post location.

All efforts to make an initial attack on the fire were abandoned within the first 30 minutes, as the Oakland companies were forced to abandon their positions to take refuge. It was during this critical period that one of the battalion chiefs was killed, along with a police officer and most of the civilians who died. The retreating units were unable to communicate effectively with the command post and had to retreat to safe areas before they could regroup and commit themselves to further action. The units that were unable to communicate attempted to establish defensive positions on their own initiative. Where it was infeasible to engage the fire, they made their best attempts to evacuate residents ahead of the flames.

As successive agencies became involved in the incident, multiple commands developed. Command posts utilized during the course of the fire and other features of incident command are illustrated on the following page. The Contra Costa County units established a secondary command structure on the east side of the fire and communicated many of their needs back to ConFire on their own radio channels. ConFire mobilized all of the resources that could be obtained from their county, in spite of the simultaneous major fire in Franklin Canyon. The Contra Costa County Assistant Chief, who was at the ConFire Center, was instrumental in obtaining the air tankers from CDF for the Oakland incident.

When the CDF units arrived, they were also unable to make contact with Oakland and established another secondary command structure, taking responsibility for the northeast flank of the fire. These forces used CDF radio channels and communicated back to Morgan Hill to obtain additional resources.

When the fire spread into the city of Berkeley, yet another independent command structure was established, with the Berkeley Fire Department units utilizing their own radio system. This meant that there were two "primary" incident commands and two "secondary" commands, working with four separate communications centers and unable to effectively communicate with each other.

The San Francisco Fire Department (SFFD) later established an additional "secondary" command post at the Claremont Hotel and coordinated its resource requirements back to the SFFD Communications Center using their own radio system.

All of the agencies involved used the ICS system in parallel, with inadvertent overlapping boundaries and designations. The first assignment in the Oakland system was to designate Battalion 4 as "Division A." After this area was abandoned (and the assigned officer was killed), Battalion 44 was designated as "Division A" on the Rockridge front. Battalion 45, directing units arriving from the Contra Costa side of the hills, designated the captain of an Orinda engine company as "Division A" on the upper east side of the fire. The Berkeley incident commander established the Tunnel/Bridge Road front as "Division A" and the Roble Road area as "Division B" by the Oakland incident commander. The two Berkeley divisions were geographically located between Oakland's "Division B" and "Division C" (Claremont Hotel) areas.

The multiple use of division designators was not a severe problem, however, since each was communicating with a different command post on different radio systems, but it illustrates the type of problems that can arise without unified command and effective communication.

At the Oakland command post there were not enough command personnel to fill standard ICS positions, so important planning functions, including resource status and situation status, had to be absorbed by the command officers who were trying to direct suppression operations. Logistics coordination was assigned to members of a ladder company. The coordination began to improve as the other agencies established liaisons at the Oakland command post.

Between 1500 and 1530 hours, a unified command structure was developed among Oakland, Berkeley, and CDF, and, by approximately 1630 hours, the fire perimeter had been divided into three branches (as illustrated on the following page) and 15 divisions, as previously noted. The unified command was centered at Oakland's command post, although Berkeley continued to operate a command post to direct operations in its area and the CDF and San Francisco secondary command posts continued to function.

The command system was further refined as the OES overhead team arrived and assisted in the implementation of a large scale command logistics system for the incident. This moved the incident into a longer-term, large-scale mode of operations that continued for the next four days.

At the onset of the fire, the incident command structure was unable to catch up with the expanding incident for several hours. Some of the major challenges included:

- The fire moved so fast and became so large that it could not be accurately defined or predicted.

- The units that were initially deployed in close proximity to the fire had to abandon their positions and evacuate the area; it was impossible to account for their welfare or direct their regrouping into effective tactical units.

- Units that were unable to communicate took actions on their own initiative and were not under the control of the command system.

- Changing conditions required frequent changes in plans.

- Communications systems could not support the necessary exchanges of information and direction of tactical resources.

- Officers assigned to survey the fire from helicopters could not communicate with the command post and had to land to deliver their information.

- Mutual aid units could not arrive fast enough to keep up with the rapidly escalating need for them.

- Incoming mutual aid units did not have maps of the fire area.

- There were insufficient command personnel to staff essential positions.

- The command post location became untenable and had to be relocated two more times.

- The final command post location was so overwhelmed with smoke and embers that it was difficult to work and impossible to make a visual assessment of the situation.

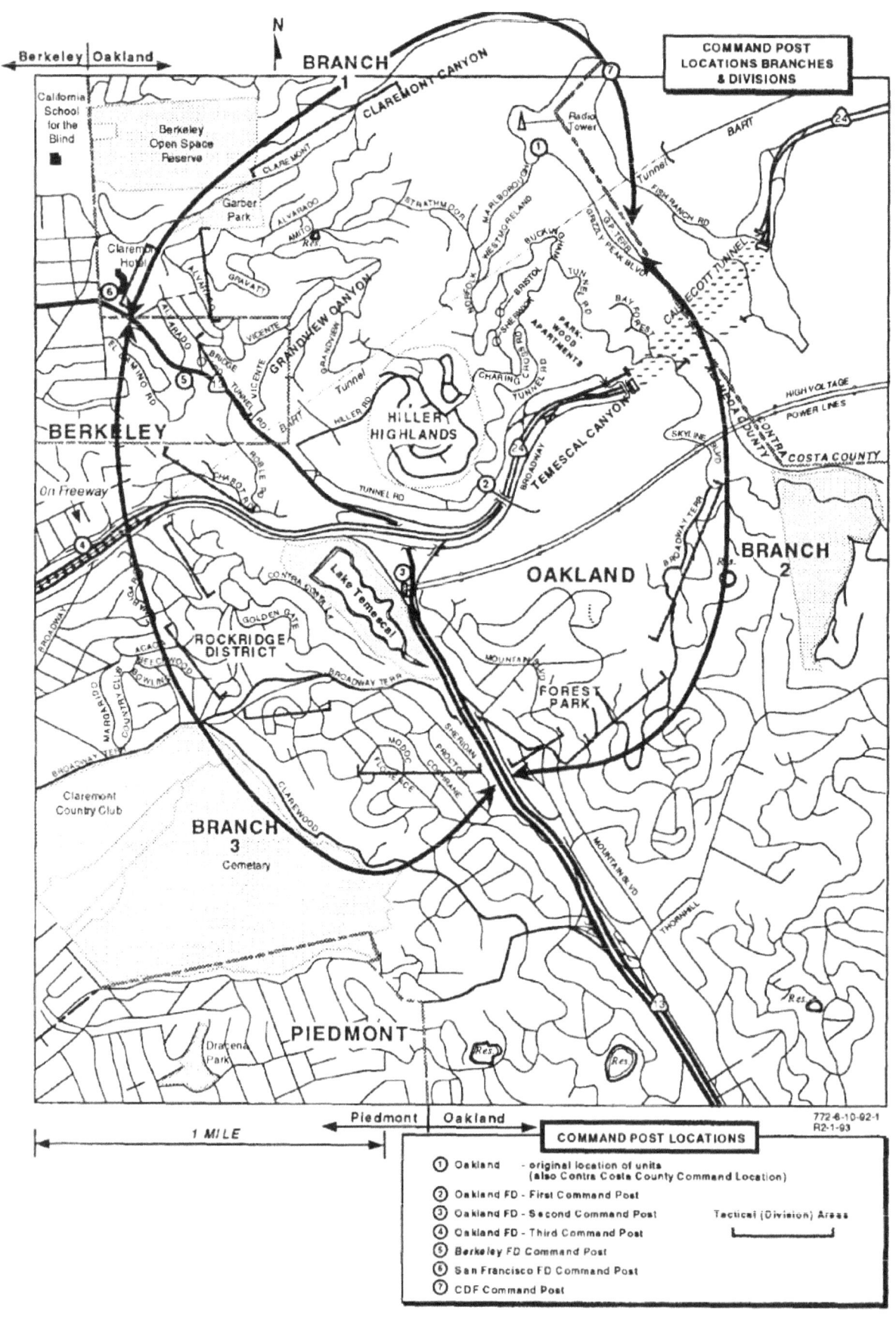

## Communications

Effective communications systems are an essential component of emergency operations. In the analysis of this incident there are several components of the established communications systems that proved to be inadequate. These points are valuable lessons in planning for future large scale situations.

*Radio* – The Oakland Fire Department conducts its emergency incidents on a single VHF radio channel, with one additional channel available for back-up. The normal mode of operations is to dispatch and manage incidents on the main channel (F2) and to keep working incidents and multiple alarms on this channel. All other traffic is switched to the back-up channel (F1), when requested by the incident commander, to reserve F2 for the major incident in progress. The digital tones for status encoders in the department's vehicles also transmit on F2, and there is an ongoing problem with the tones overriding voice transmissions. There are also locations in the hill areas where radio reception is difficult because of signal blockage caused by the terrain.

The Oakland Fire Department's mobile and portable radios also carry White 1, the primary Statewide mutual aid channel, which is used for interagency communications and as a back-up for the other channels. Command officers have White 2 in their portable radios and use it as a command channel during major incidents. The third Statewide channel, White 3, is not installed in Oakland's radios.

On the day of the Hills fire, all six alarms of Oakland units were on F2, placing an extremely heavy message load on the channel, in addition to numerous status tone transmissions as each unit reported it was responding and then on the scene. When numerous units are attempting to transmit status information simultaneously, they begin to block each other, causing the devices to retransmit the message until an acknowledgement is received from the central processing unit. Each transmission occupies the channel for only a fraction of a second, but each interruption makes voice communications more difficult.

With the arriving units requesting instructions and units on the scene reporting rapidly changing fire conditions, the contention for air time was severe. It became very difficult to transmit a message to an intended receiver and to obtain an acknowledgement without being interrupted. Units with urgent messages had to wait for opportunities to talk.

When the command officers tried to go to White 2 or F1 to communicate with each other, they removed themselves from the critical traffic on F2 and were out of touch with the communications center. They had to switch back and forth to try to communicate with each other and with their companies. Traffic relating to other incidents was also occupying F1.

By the time the units on the upper slopes had to abandon their positions, unit to unit communications had become extremely fragmented. The incident commander had no means to keep track of the location, status, or welfare of units and could not be sure which units were safe or in trouble. Companies were urgently requesting assistance, but there were no units to send and no means to direct units to where the help was needed. Without effective communications, it became an undirected and uncoordinated situation, with companies doing whatever they could to provide for their own safety and evacuate residents in the path of the fire.

It was during this period that the battalion chief was lost. His last radio transmissions indicated that the fire had overrun the positions on Westmoorland Drive, and he was pulling the units out of the area. Moments earlier, the operations assistant chief had expressed his concern for the danger to firefighting personnel, advising him "Don't get anyone killed up there." Sometime after that message was transmitted, Battalion Chief Riley was killed, but in the confusion he was not missed until his car and then his body were found by units reentering the area, more than two hours later. (The radio tape indicates that he may have tried unsuccessfully to communicate as late as 1222 hours, approximately 30 minutes after his last successful communication.)

The Oakland command post vehicle was initially dispatched to the top of the hill at Grizzly Peak and missed the first instruction to set up at the tunnel entrance; it had to be redirected to the bottom on Highway 24. Until it arrived, the incident commander had only a portable radio. Soon after the vehicle arrived and was set up, it had to be relocated, with the fire on the verge of overrunning its position. This created a significant delay in setting up the command post with effective communications and support systems.

When they were unable to communicate by radio, company officers resorted to other means. Some messages were transmitted by police officers to their communications center, then relayed to the fire department. Others called in on private telephones from houses or on cellular telephones borrowed from spectators. Messages for the incident commander had to be relayed back to the command post from the communications center. Cellular telephones provided reliable auxiliary communications for several functions.

White 1 was in use by the EBRPD and Oakland units that were originally on the scene of the fire and was the designated channel for units arriving on mutual aid to report in. The first arriving Contra Costa County units were unable to make contact on White 1, so they continued to communicate with each other on Contra Costa channels, while the battalion chief went to report in person at the command post. Contra Costa County's White 1 base station was disabled when fire reached the telephone lines to the transmitter site.

As additional mutual aid resources approached the scene, they were also unable to make contact with the command post for directions and assignments. These units either found the command post and reported in person or went into action where they encountered the fire. Lacking detailed maps of Oakland city streets, they also had difficulty navigating to assigned destinations, and several found themselves in the middle of suppression operations, but in the wrong locations. Meanwhile, division supervisors who had requested reinforcements received none.

White 1 was also designated for helicopters to communicate with the command post. Eagle 5, the EBRPD helicopter, was over the scene, but could not get through to the command post to provide information on fire growth and spread. The battalion chief, who was assigned to size-up conditions from a police helicopter, had to land to give his report. Copter 106 had similar problems and had to make its own determinations of where to land and discharge its crew members and where to attack the fire with water drops. A captain at the Oakland Fire Department Communications Center used White 1 to coordinate the aircraft operations for part of the incident but experienced difficulty coordinating needs with the command post.

Berkeley, San Francisco, and CDF resources all used their own radio channels, supplemented by cellular telephones, to communicate with their own units and with their respective communications centers. The ability of the other major agencies to utilize their own radio channels actually relieved some of the pressure on the main command post and radio channels. As long as effective liaisons could be maintained, this was the most efficient means to reduce the communications overload at Oakland's command post.

The command post was overwhelmed by communications coming from too many directions, with two few people to manage. The pace was fast and furious and the overload on radio channels prevented effective communications from taking place. To utilize a radio channel effectively, there should be one person at the command post assigned exclusively to communicate on that channel. That individual must be able to input significant information to the decisionmaking process and obtain direction.

The radio problems began to improve when two radio caches were delivered to the scene from the Alameda County Coordination Center. Each cache included 40 multi-channel portable radios and a mobile repeater unit. This additional radio capability, along with sufficient personnel to staff an expanded command post, provided relief for the critical radio communications problems by approximately 1600 hours.

Oakland has ordered and is awaiting delivery and installation of an 800mHz trunked radio system for the fire department. This system will provide a needed improvement in the availability of radio paths and will counteract some of the overload problems on the existing channels. However, this will not solve the problem of too much information coming too quickly into the command post, with too few personnel to perform necessary command functions and effectively communicate with units at different locations. There is an opportunity to establish a multi-jurisdictional trunked radio network among several of the jurisdictions in the North Zone of Alameda County. This could result in a major operational improvement for the mutual aid system.

**Communications Centers** – The Oakland Fire Communications Center was at minimum staffing on the morning of the fire, with two operators and one supervisor on-duty. The center, located in Fire Station One, has seven consoles, each capable of performing all radio and telephone functions. A new, second generation computer aided dispatch system had recently been installed in the center.

The operational policy is for individual operators to answer incoming telephone calls, dispatch incidents, and handle the radio traffic associated with their incidents. This means that each operator may have to deal with both telephone and radio traffic at the same time. The Oakland Police Communications Center initially answers all 9-1-1 calls and transfer fire-related calls to the Oakland Fire Department Communications Center.

As soon as the fire broke out, the fire communications center was flooded with incoming calls from citizens reporting the fire and providing information, and news media requesting details. As the fire grew rapidly, the calls changed to requests for information on whether or not residents should evacuate and distressed residents reporting that the fire appeared to be heading toward them. Some residents asked for instructions on which way to go to get away from the fire. These calls kept all of the incoming telephones ringing incessantly, keeping the operators on the telephones and making it difficult for them to monitor radio traffic. The supervisor was occupied with making notifications to designated individuals and agencies on the multiple alarms and requesting mutual aid from adjoining jurisdictions.

With telephone traffic occupying the communications operators, it became difficult for the incident commander or any units at the incident scene to communicate effectively with the communications center by radio. Messages going in both directions had to be repeated and miscommunications became a major problem. Cellular telephones were used as an alternate means to call in to the communications center, but they too faced the problem of too many busy telephone lines and not enough personnel to answer them.

The overload on the operators resulted in delays in processing the requests for multiple alarms, mutual aid, and other requirements. Outbound calls had to wait because there were no lines available to call out of the communications center. Messages and requests were backed up and confusion increased.

Part of the delay in sending air tankers was attributed to a lack of familiarity with standard terminology by the person who made the call to CDF. When CDF called back to clarify the request for assistance, the call was put on hold because so many telephones were ringing and every message appeared to be critical. It was impossible to sort out who was on which line and what was most important at any moment. It was approximately 15 minutes from the time the incident commander requested assistance from CDF until Copter 106 was dispatched and 25 minutes before CDF ground units were en route.

Within the first 15 to 20 minutes, additional personnel arrived at the communications center to assist the on-duty shift. Several trained operators, officers, and firefighters arrived and activated the additional consoles, but they were not able to relieve the overload condition and the communications center became severely congested. It was impossible for the operators to comprehend the situation that was going on at the scene, so they could not provide evacuation instructions to the residents seeking direction.

The communications center handled hundreds of requests from the command post throughout the day, and the operators struggled for hours to support the operation. The command post was calling with all kinds of different urgent requests, and other agencies were calling to ask if they could help or to advise that they had special equipment available. It was a massive overload situation for the personnel and the systems that were available to them.

Berkeley operates a much smaller public safety communications center which had one 9-1-1 call taker and one fire dispatcher on-duty on the day of the fire. During the first hour they received dozens of calls from concerned citizens about the fire and reassured them that Oakland was taking care of it. When the fire unexpectedly attacked Berkeley, the communications center was suddenly placed in a similar situation to the Oakland Communications Center. One of the problems that was reported in Berkeley was that the person answering the telephone continued to reassure callers, well after the Berkeley incident commander had ordered a massive evacuation of the area ahead of the fire front.

ConFire was in the difficult position of trying to handle multiple major incidents at the same time. The Franklin Canyon fire was demanding multiple alarms at the same time as the East Bay Hills fire; one in the county and one on the border threatening to come into the county. ConFire was able to send multiple alarm assignments to both fires, called up additional strike teams from other parts of the county, and then called for mutual aid from Region II to cover the massive void of coverage in Contra Costa County. It then had to deal with two more major fires in Contra Costa County, when the power lines shorted out.

ConFire continued to manage resources among all of these incidents throughout the day and continued to find more resources to send to Oakland as the other fires were brought under control. ConFire provided communications support for the resources it sent to Oakland, since they could not establish effective communications with the Oakland Command Post, and also served an important role in obtaining the air support for Oakland when the need was critical. (A description of activities within Contra Costa County is included in Appendix C.)

## Public Information

Notification of the public becomes a very significant concern in any disaster situation, when there is a need for evacuations or other actions by the general population. The established system in Oakland calls for the city's public information staff to establish a central emergency operations/information center to disseminate information, primarily through the news media. Since the emergency occurred on a Sunday, the public information staff had to be called in from home and took a considerable amount of time to gather the information and to be prepared to perform their functions.

While this delay was occurring, most of the news media had representatives on the scene, covering the story and giving out whatever information they could obtain from any sources that were available. In several cases it was suggested that the media obtained better information from unofficial sources than from the "official sources." This causes confusion and may cause improper actions to be taken by the public.

There was an additional problem with law enforcement agencies excluding the news media from operational areas. This caused conflict with media representatives on the scene who were aware of their rights to enter to gather legitimate news information.

It should be emphasized that public information is a critical component in a disaster, particularly if the public must be warned to take protective or preventive actions or to evacuate an area that is in imminent danger. There should be procedures to provide for rapid activation of the public information system and close coordination with the incident commander.

## Mutual Aid

The established mutual aid systems among cities, within counties, and on the regional and Statewide levels provided an unprecedented amount of assistance to Oakland and Berkeley for this incident. The total mutual aid commitment involved 440 engine companies and 1,539 personnel from 250 agencies. Eighty-eight strike teams were mobilized, some from more than 350 miles away, and some were on the scene for four to five days.

The effectiveness of this mutual aid system in being able to mobilize and support this level of response is unparalleled in the United States. The analysis of the events indicates some areas where procedures could be improved to speed response and improve coordination, but these suggestions are based on a situation that exceeded all previous experience.

The State of California Mutual Aid Plan was based primarily on experience gained from wildland fires, some of which became wildland-urban interface fires. The procedural steps that are involved in this plan do not facilitate rapid response. In designing the system, extremely rapid response was considered to be less essential than establishing and maintaining strong organizational characteristics. Specific requests are processed from the requesting jurisdiction, to the County Coordinator, to the Region, and, if necessary, from one Region to another. The strike teams are then activated in a very structured manner, with mobilization orders passed down through the system to each individual fire department that will be contributing a unit to a strike team. The departments are notified to have their units form up in convoys at designated meeting points.

These very deliberate procedures reinforce the system, but they take time. The procedures for notification may take 15 to 30 minutes, particularly if the request is for several strike teams. It is not unusual for a strike team to take an hour or more to assemble and be ready to depart for its destination, which may involve several hours' travel time. The delay is not a problem in cases where resources are being assembled in preparation for strategic actions, but it is a significant concern when a conflagration is moving through an urban area and lives are being lost.

The Bay Area Inter-County Mutual Aid Plan is designed for rapid response within the highly urbanized six county area. It took six minutes for the request from the incident commander to be transmitted from the Oakland Communications Center to the San Francisco Fire Department Communications Center, then it took only 90 seconds for the dispatch message to be transmitted to the units that would form the first strike team. The second strike team had assembled and was en route to Oakland within 11 minutes of the call. The response from Contra Costa County was also immediate, dispatching a first and second alarm in rapid succession, then calling up additional strike teams without waiting for requests to work their way through the formal system.

The Alameda County Fire Mutual Aid Plan is also designed for rapid response but encountered problems on the day of the fire. The interconnect line among the departments in the North Zone was not operational and calls had to be made from Oakland to each individual neighboring department to request assistance. In the confusion of the situation at the communications center, Oakland did not pass the zone coordination responsibility to Alameda Naval Air Station (NAS) and neglected to notify or request assistance from Berkeley.

At the Alameda County coordination level, the processing of the initial request took several minutes, because of the procedural requirements and conflicting radio and telephone traffic. Recognizing the problems, steps were immediately implemented to speed up the process for the requests that followed. It appears that there are improvements that could be made to process intra-zone and inter-zone requests more promptly by pre-designating responses and developing an immediate notification and dispatch system. The situation was more complex, because some mutual aid requests were handled directly, by calling to other jurisdictions, while others went through the county coordinator.

The procedures for requesting CDF response also appear to be a problem, particularly when aircraft are needed. A major priority is directed toward determining if the fire is in a "State responsibility area" or a "threat zone" where CDF would assume the cost of suppression, or if the local jurisdiction will be responsible for the cost. If it is a local jurisdiction fire, a specific request for air tankers is required.

There were difficulties in communication between the Oakland Communications Center and CDF on the initial request for assistance, which brought the first helitack unit. There seemed to be a problem convincing CDF of the urgent need for additional aircraft at the East Bay Hills fire, even with Contra Costa County recommending diversion of the air tankers from the Franklin Canyon fire to Oakland. The personnel in the Oakland Communications Center believed that air tankers and more helicopters were on the way at least 30 minutes before CDF rerouted the first two air tankers and dispatched a third.

The procedures for requesting CDF response, authorizing aircraft response, and coordinating air operations need to be reexamined and streamlined for urgent situations. Time is much more critical in a wildland-urban interface fire than in a normal wildland situation.

## Volunteer Response

While the large scale mobilization of fire department resources was taking place, off-duty firefighters from Oakland and surrounding communities were converging on the scene to offer their assistance.

Appeals were broadcast for all available experienced firefighters to respond to Oakland to assist in stopping the conflagration, and scores responded.

Many individual firefighters, both career and volunteer, simply showed up, attached themselves to operating units, and went to work. They were joined by dozens of citizens who pitched in to help the firefighters in any way they could. The untrained citizens carried equipment, dragged hoselines, helped to direct nozzles, and performed anything that was asked of them. There were also groups of military personnel, particularly from the Alameda Naval Air Station, who showed up and volunteered their services. While trained firefighters wearing protective clothing worked in the most hazardous locations, they were often supported by citizens in shorts and tee shirts who were anxious to help and enthusiastically followed directions.

While untrained citizens were being enlisted at the fire scene, trained firefighters from other jurisdictions who called to offer their services were directed first to Oakland Fire Headquarters and later to the staging area at Raimondi Park. Dozens responded and signed in to offer their services, but there was no mechanism to organize them and utilize their capabilities. Some complained about being kept waiting around the staging area for hours, watching reports on television and looking at the fire in the distance. Some became frustrated and drove to the fire scene on their own, joining in the action wherever they saw the need.

There is a dilemma in the concern for accountability and safety of "spontaneous volunteers," versus the desirability of using them to assist in situations that are beyond the control of the regular force. Clearly, untrained citizens provided valuable assistance in several areas. Oakland had a very similar experience with citizens assisting in the initial rescue efforts at the Cypress Freeway collapse in the Loma Prieta earthquake. In disaster planning, there is a need to develop mechanisms to effectively utilize this "spontaneous volunteer" assistance.

There must also be concern for the safety of such participants. Prevailing standards hold the incident commander and supervisory officers responsible for the safety of all personnel under their direction. This should be a consideration when the use of untrained personnel is contemplated, but in a true disaster the decisions are usually made at the point where the need exists, not at a central command post.

Prevailing standards and procedures severely limit the ability of trained firefighters to "freelance" at incidents, by requiring them to be part of a recognized organization structure. It is ironic when untrained citizens can be used, but trained and capable firefighters cannot, because there is no structure to bring them into the system. While the need for such mechanisms may be infrequent, there should be a system to effectively organize both trained and untrained individuals when the need arises. This should be addressed in planning for future disasters.

## Aircraft Operations

The need for aircraft to attack the fire was expressed from the first stages of the incident. The captain of Engine 19 requested mutual aid from CDF at 1059 hours, and this was echoed by the assistant chief who was still en route at 1104 hours. Both of these officers expected that CDF would be responding with air tankers to drop slurry on the flanks of the fire, as well as helicopters to make direct water drops. The assistant chief used the key phrase "threat zone" to indicate that the fire was in an area where it could spread to wildland areas within CDF jurisdiction.

When the request was passed from Oakland Fire Communications to CDF at Morgan Hill, there was confusion over the terminology that was used. The normal mutual aid response, which includes a single Helitack unit, was dispatched. The confusion was based on the terminology of the request for "air operations," since this refers to a particular individual in the ICS structure. It took several minutes to clarify that the request was for air support and the only available air tankers had been dispatched to other incidents.

When the officers at the scene made repeated urgent requests for aircraft and were reassured that they were coming, only the one Helitack unit was actually en route. It was not until 1236 hours that the two tankers responding to the Franklin Canyon fire were rerouted to Oakland, and it was 1300 hours before they were overhead. A third air tanker was dispatched from Ukiah at 1239, with an estimated time of arrival of 1316 hours.

The diversion of the first two tankers and the dispatch of the third were based on the intervention of the Contra Costa County assistant chief, who convinced CDF of the urgent need for aircraft and accepted responsibility for the cost. Several more air tankers were deployed to Oakland during the afternoon, as quickly as they could be released from other wildland fires in California. Contra Costa County also activated a privately owned (contract) water drop helicopter to assist Oakland.

Initially, a captain at the Oakland Communications Center tried to coordinate the aircraft operations over the radio. Later in the afternoon, a CDF battalion chief, located at the command post, took over the task of communicating with the spotter aircraft and communicated the priorities for air attack operations. The primary objectives were to stop the fire's advance to the southwest, along both sides of Highway 13. The area where the Berkeley Fire Department was trying to make a stand, north of Highway 24, was a third priority for the aircraft.

It was extremely difficult for the spotter aircraft to see through the smoke and for the tanker pilots to find a clear run to their target areas. To make bombing runs into the wind would have required the aircraft to fly directly toward the smoke-obscured hills. Flying with the wind required the aircraft to maintain higher than normal altitudes, and the combination of altitude, wind, and thermal updrafts significantly reduced the effectiveness of their drops. After each run, the aircraft had to return to Santa Rosa Airport to refill their slurry tanks; a round trip which takes approximately one hour.

Water drops from helicopters were more effective in reaching their targets, but their water capacity is limited and the pilots had great difficulty with the wind and smoke conditions. Lake Temescal provided an ideal source for refilling the helicopter water buckets, within 60 seconds flying time of the drop zones, and Lake Merritt was available as an alternate. The helicopter drops are most effective in quenching a particular hot spot, as opposed to the fixed wing tankers' specialty of dropping slurry along a line. The fire's thermal column was extremely hazardous to low flying helicopters, forcing the pilots to work carefully along the flanks, avoiding the head of the fire where the superheated air would have been disastrous.

**Water Supply**

Water supply was a major problem during most of the incident. Part of the problem related to the fact that many of the units that responded from distant areas were unable to hook up to Oakland hydrants. When California adopted a standard 2-1/2-inch threaded connection for all hydrants, the cities of Oakland and San Francisco opted to maintain their 3-inch connections and to keep a supply of adapters on hand for mutual aid units. Fire departments in the area normally carry adapters on their apparatus, but the plan called for adapters to be obtained from the warehouse to meet incoming mutual aid strike teams at staging areas. Since this fire occurred on a Sunday, there was a delay in obtaining the adapters until off-duty personnel could open the warehouse and send them to the scene on supply trucks.

Many of the incoming units were committed and discovered the adapter problem only when they needed water to supply hoselines or refill their tanks. This limited the ability of several units to work effectively until they could locate a unit with an adapter, or one of the supply trucks located them. Since some of these companies were in critical combat areas, it was difficult for the logistics system to find them and deliver the adapters.

The water supply on the hills was known to be a problem from previous incidents and from risk analysis projects, including earthquake vulnerability studies. The water system on the hills was arranged as layered pressure zones, each supplied by a tank at a higher level. The storage tanks served areas where the difference in elevation would maintain static pressure in a desirable range at the delivery levels.

The tanks were kept filled by a serious of electrically powered pumps, which relayed the water from tank to tank, and the pumps were not provided with emergency generators. If a pump at a particular level failed, it isolated the tanks at higher levels from any capability for replenishment. The power began to fail early in the fire, as wooden poles burned, lines dropped, and transformers exploded. As pumps failed, the higher level tanks would begin to run out of water. When the high voltage lines shorted out, at 1315 hours, all of the power to the remaining pumps failed, and the whole system on the hills began to run dry.

The demand on the system was also very high, as companies tried to establish large handlines and master streams to establish defensive lines. In addition, many of the homeowners were using their garden hoses to wet down their roofs and shrubbery to guard against flying brands and embers; some even left garden sprinklers running on their rooftops as they evacuated. As homes burned to the ground, their water connections were left spurting water into the rubble. All of these factors created an unprecedented demand on the system, quickly using all of the stored water.

Companies on the hills reported hydrants going dry as early as Sunday noon, and the supply was not restored until that night, when portable generators were brought in to power some of the critical pumps. It does not appear that the water supply was a deciding factor in the outcome of the fire on the hills, since the crews were unable to make any progress against the flames before the hydrants went dry. The strength of the wind and the thermal forces made water almost totally ineffective to stop the downwind progress of the fire. The available water was useful in protecting certain positions, including some locations where firefighters took refuel, and in covering exposures on the flanks.

In the Rockridge district there were also sections where the water supply was known from past experiences to be weak. Many of the mains in the area were old and inadequate, and at least 50 homes were burning by 1300 hours. San Francisco Strike Team One was assigned to this area and around 1420 hours the strike team leader was able to call back to his department and have two of the city's large diameter hose tenders activated and dispatched to Oakland. The hose tenders were able to bring in large supply lines from streets on the edge of the district to supplement the supply.

One of the strong water supply areas was the private system installed at the Claremont Hotel. This system provided an adequate supply for the defensive streams that were established on the exposed side of the hotel. While these streams were maintained in a stand-by defensive posture, the crews were able to extend handlines up the hill to engage the fire on Alvarado Road and some of the smaller streets overlooking the hotel. This kept the fire from advancing further down the hill and causing a direct exposure to the hotel.

## Stress

Both physiological and psychological stress had significant effects on the firefighters who were involved in this incident. The physiological stress aspects are easy to comprehend, since many firefighters were involved in continuous and intense fire suppression activities for hours on end. Most of the on-duty Oakland crews and the mutual aid forces that responded in the first two hours were actively engaged in fire suppression for 12 hours or more, without more than a few minutes of rest at any time.

The reductions in the Oakland and Berkeley Fire Departments' budgets had reduced their staffing levels and number of companies and made their challenges even more difficult in this incident. While it may be assumed that reduced company staffing subjects the crew members to additional stress and exertion, it is clear that every firefighter on the front lines was subjected to a maximum amount of stress and exertion at this fire.

The heat of the fire and the exertion of extended heavy labor drained strength and challenged the stamina of firefighters well beyond normal limits. Even when they could take a brief rest, the 900 ambient temperatures and the strong hot wind provided no respite. Officers had to be extremely concerned with the dangers of heat exhaustion, dehydration, and simple fatigue from overexertion. There were few opportunities to rehabilitate or rotate crews for most of the day and providing drinking water and fluid replacement drinks for the front line combatants was a major problem. After several hours, volunteer groups, including the Red Cross and Salvation Army, began to deliver food and drinks to the suppression forces wherever they could be reached.

It would be reasonable to conclude that many of the participants pushed themselves far beyond their normal limits of stamina and endurance, based on the magnitude and potential consequences of the situation. As long as the fire did not quit, they were determined to keep up the battle. But there was a definite psychological impact of being unable to stop the enemy as it overran fire department defensive lines time and time again. In the past they had always been able to attack and defeat the fire or as least define and hold a perimeter around it; this fire could not be defeated and it could not even be constrained.

There were a few reports of conflict between units regarding the tactics employed and the level of effort that was being directed toward operations. These seemed primarily to involve units that were extremely fatigued and had reached the point where their physical stamina and aggressive spirit had both been depleted, encountering units that were fresh or supervised by unrelenting officers. These situations were unusual, but they were widely reported, suggesting that conflict existed between different agencies or between structure-oriented units and wildland-oriented units. The more frequent observation was that everyone worked together very effectively, in spite of the many adversities created by the fire itself.

Another psychological factor was the fear of being overwhelmed by the fire. Few of the firefighters had ever imagined an urban conflagration of this magnitude, and no one had ever seen one. Many experienced firefighters expressed their feelings of fear as the fire jumped from building to building, swept over and around them, and overwhelmed their efforts to control it. They described their feelings as the smoke reduced their visibility to zero, and they wondered if the fire was going to close in on them or cut off their escape.

Many of the companies that had been involved in the initial actions on the hills had been driven out by the fire, abandoning their positions and, in some cases, losing track of crew members. One company officer did not know the fate of his crew for several hours; he drove the engine down from the hills alone. There was no effective organization structure in place, and the communications system was so overwhelmed that it was impossible to account for companies or individuals. The battalion chief who died was not missed in the confusion until his body was found, and there were several reports about finding burned bodies in the streets or not being able to rescue persons who were known to be trapped in burned homes. This information spread among the firefighters on the scene and helped to create an extreme stress level.

Many of the individuals who were involved in fighting this fire were also involved in the Loma Prieta Earthquake that struck the Bay area in 1989. The Cypress Freeway collapse in Oakland had been a devastating incident, in terms of its magnitude and consequences, and most of the crews that were involved in the first few hours of that incident were also on duty on the day of the fire. They described the fire as being more stressful, because of the continuing fear that the fire kept getting bigger and nothing they could do seemed to stop it. While the earthquake was an extremely stressful incident, they did not feel a personal commitment to stop a freeway from falling down. They did feel a personal commitment to control fires, but they were facing a fire that was a major threat to their own safety.

The Oakland Fire Department instituted its Critical Incident Stress Debriefing system in the days immediately after the fire. The long term consequences of the stress are not known, but it was evident that a very high level of stress surrounded the incident, particularly the loss of so many lives and such a large part of the community. The mental health agencies and professionals in Oakland and Berkeley also provided counseling for residents of the devastated area. The sight of more than two square miles of devastation, with the total destruction of more than 3,000 dwellings, is convincing evidence of the need for these services.

Some of the most difficult feelings were faced by command officers who were responsible for managing and directing the fire suppression efforts. There were experienced officers who were extremely familiar with the risk factors and the history of previous fires in the hills and had some expectation of what could happen, but the actual fire was worse than their worst predictions. Some of these individuals had tried unsuccessfully for years to make residents and political leaders aware of the risks and to convince them to take measures to mitigate the hazards. They had a high level of personal commitment to prevent this fire from happening and to be prepared to manage the situation if it did happen, but they were defeated in both respects by forces beyond their control.

It was particularly devastating when reports were published after the fire accusing the fire department of negligence or inadequate response to the fire. Some of those accusations came from fire service professionals who had not been involved in the incident and were not aware of the situation. These accusations caused additional stress to several experienced and conscientious officers, who were already suffering from physical and psychological stress.

## Firefighter Safety

Approximately 150 firefighter injuries were reported on this incident, plus the one fatality. Most of the injuries were in the categories normally experienced in major fires, including exhaustion, dehydration, strains, sprains, contusions, dislocations, and minor burns. Several cases relating to smoke inhalation and eye irritations were also treated, and there were numerous reports of post-incident exhaustion, as indicated by weakness and aching muscles. At least one firefighter with chest pains was handled by the medical units at the scene. Another was blown off a roof by the force of the wind.

Many of the firefighters were treated for respiratory distress, eye irritation, and other symptoms of prolonged exposure to products of combustion, exacerbated by dehydration, heat exhaustion, and stress. Treatment was provided at first aid stations, in rehab areas, and by ambulance crews that were assigned to the incident. Several firefighters were treated at hospitals, and some were held overnight for observation and additional treatment.

There was a problem in trying to decide if this incident should be handled as a wildland or structural fire. It primarily involved exterior operations but also included most of the hazards of structural operations, including search and rescue and interior attack on several structures. Many units alternated between the two types of activities, working mainly on the exterior but engaging in structural actions whenever it was the most effective way to deal with a situation.

The accepted protective equipment for a wildland intervention includes a lightweight fire protective jacket and pants, with a helmet, goggles, leather boots, and gloves. Structural protective equipment includes insulated coat and pants, helmet with ear protection or hood, gloves, rubber or leather boots, and self-contained breathing apparatus. Some of the responding units had only one type of protection or the other, while some brought both.

The use of structural firefighting protective clothing for hours on end, on a hot day and in a fire's thermal environment, creates an obvious problem with fatigue and heat exhaustion. The addition of breathing apparatus provides protection for the respiratory system and eyes, but adds more weight and requires regular changes or refilling of the air cylinder, which would be infeasible on the scale of this incident. The wildland protective clothing is more practical for operations that extend over lengthy periods, but does not provide the same level of safety, particularly respiratory protection.

There is no easy middle ground between structural and wildland protective clothing. The individuals involved in the incident used whichever they had with them, although most expressed a preference for wildland clothing under similar conditions.

The lack of respiratory protection was a major concern, in both interior and exterior operations, and there were numerous reports of respiratory irritation. Fortunately, there were no major injuries related to the lack of proper protective clothing or respiratory equipment. This is a subject that needs further research, to determine the best protective ensembles and tactical approaches for an interface fire.

A few of the crews were wearing experimental respirators (smoke masks) that have been developed by Lawrence Livermore Laboratory for use by wildland firefighters and are currently being evaluated in California. The masks incorporate a full face piece with a large capacity HEPA (high efficiency particulate) filter intended to remove soot, ash, cinders, tar, and other airborne contaminants from the inhaled air. The filters do not remove carbon monoxide or other gaseous contaminants.

After several hours of use, the team members all registered very high on carboxyhemoglobin (the amount of carbon monoxide in the bloodstream), indicating that they had been exposed to significant concentrations of carbon monoxide. The members who wore the masks reported that they were able to rest and recover from the carbon monoxide saturation in a relatively short time, while those who worked without the masks had much more respiratory action and at least two required hospitalization.

This experience provided valuable information on two points of interest. One finding confirmed that members working "in the open" many still be exposed to high levels of carbon monoxide from wildland fires. The second is that the other combustion products and contaminants play a major role in causing severe respiratory reactions and may even increase the toxicity of carbon monoxide. This synergistic effort suggests that a filter-type mask may prove to be valuable and effective in wildland fire suppression.

The experimental masks currently incorporate a monitoring device that indicates high carbon monoxide levels in the environment, as a warning to discontinue work in that area. Future developments of the mask may incorporate chemical filter packs to remove carbon monoxide from the air, similar to the filter canister masks that were once popular for structural fire suppression but proved to be unsafe in oxygen deficient atmospheres.

The safety officer position was not staffed until late on Sunday afternoon because of the multitude of problems at the command level. This incident involved extreme risks to firefighters, and to adequately address the responsibilities of the safety officer position would have required numerous qualified individuals maintaining communications with a senior safety officer at the command post.

## Emergency Medical Services (EMS)

The primary agency for emergency medical incidents in Oakland is the Alameda County EMS, which operates advanced life support ambulances in Oakland and some other parts of the county. The Oakland Fire Department provides first responder service to certain categories of emergency medical incidents. In some jurisdictions within the county, fire departments provide advanced life support and ambulance transportation.

When the third alarm was transmitted, Oakland Fire Communications advised the EMS Communications Center that the fire department was suspending its response to medical incidents, according to established policy. During the fire Alameda County EMS took responsibility for all medical responses at 1107 hours. As the magnitude of the incident became known, additional ambulances were activated and units from surrounding jurisdictions and from ambulance companies were placed on standby.

At approximately 1330 hours the first reports of fatalities and serious injuries were received, and the system moved to a higher state of preparedness, with several ambulances staged around the fire area. Hospitals and hospital-based helicopters were also alerted. A number of civilians with serious burn injuries were transported in the early part of the afternoon.

Alameda County EMS provides advanced life support ambulance service in the city of Oakland, and managed the treatment and transportation of injured firefighters and civilians during the fire[3] Alameda County EMS requested assistance from surrounding areas, including the city of Alameda

---

[3]The Oakland Fire Department normally provides first responder level treatment on selected EMS calls, usually arriving several minutes ahead of an ambulance. When multiple alarm incidents reduce the department's resources, the department suspends its response to EMS incidents and returns full responsibility to the County EMS agency. This results in longer response times, but there is no documentation of any positive or negative consequences resulting from this.

Fire Department and the Berkeley Fire Department, which operates advanced life support ambulances, and from private ambulance companies. All available units were placed in-service and staged in several locations around the fire perimeter. They were directed to the locations where assistance was needed by a liaison at the main command post. Hospital-based medical helicopters were also used to transport critical patients.

The ambulance crews provided medical evaluation and treatment at rehab areas and responded to locations where injuries were reported. The total numbers of injured firefighters who were treated or transported is believed to be in the vicinity of one hundred. Several additional patients were transported to areas by other means, some injured by the fire or smoke and many suffering from anxiety from the loss of their homes and family members.

For the next 10 to 12 hours the ambulances responded to treat numerous firefighters suffering from minor burns, cinders in eyes, respiratory irritation, heat exhaustion, and a variety of other injuries. Most of the firefighters were treated on the scene and returned to duty on the fire lines. Several civilians were also injured assisting in fighting the fire or evacuating from it and several cases of stress were reported, involving evacuees and spectators. The major problem for ambulance personnel was finding the areas where firefighters were working and making access to reach them.

## Evacuation

The fire burned along the north face of Temescal Canyon faster than any kind of organized evacuation could be implemented. The fire swept through areas, particularly the Hiller Highlands complex, and residents had to flee as the fire ignited around them; several died in the process. Firefighters attempted to direct residents out of the fire's path, in some cases as they themselves were retreating from the advancing flames. Police officers were also dispatched to the area to assist in warning and evacuating residents in the path of the fire. The police officer and the fire department battalion chief who died in the fire were both attempting to rescue residents and were overrun by the fire in the narrow streets of the hillsides.

As more resources were assembled and the rate of spread of the fire decreased, there was more time to plan and organize evacuations from areas in its path. Several of the flatland areas were planned for evacuation, and emergency shelters were opened to accommodate the displaced occupants.

The population of the areas evacuated on the day of the fire is estimated in the range of 20,000 to 30,000. Many of these residents were unable to return to their homes for several days due to the continuing fire threat, as well as interruptions in electrical, gas, and water service and blocked streets. There were some reports of looting in the evacuated areas, in spite of a heavy commitment of law enforcement personnel to secure the perimeter.

With over 3,000 dwelling units destroyed or heavily damaged, finding shelter and providing emergency assistance for the homeless was a major concern. These functions were assigned to the local emergency preparedness agencies, assisted by the California OES, and Federal Emergency Management Agency (FEMA) personnel.

## LESSONS LEARNED

There are hundreds of lessons to be learned from this incident, many of which are contained in the body of this report and in several other reports that have been prepared on the incident. The lessons revolve around specific themes:

**Mitigation** – The risk of disastrous wildland-urban interface fires has been recognized and emphasized by the fire service and other organizations for many years. This incident is the ultimate experience, to date, of those dire predictions coming true.

The factors that created the critical fire risk situation in the East Bay Hills on October 20, 1991, exist in hundreds of other locations and, when the same circumstances repeat themselves, there is every reason to expect that another very similar fire will result.

In particular, in California, when the Santa Ana (or Diablo) wind is blowing, and a fire occurs in a susceptible area, there is very little that any current fire suppression forces or technologies can do to resist the spread of the fire. The results will depend mainly on the fuel that is downwind from the fire and the length of time that the wind continues to push the fire in that direction.

An observation was made that most fire codes would have required an area with these fire risk characteristics to be evacuated. It was compared to a neighborhood with spilled gasoline flowing in the gutters. This type of regulatory control has never been applied in an interface area in the same way it is routinely applied to structures. The outcome of this incident appears to demonstrate the validity of the concept.

Several of the risk factors that make an area susceptible to an interface fire can be mitigated, to reduce the level of risk:

- Use of drought-tolerant and fire-resistant landscaping.

- Fuel control measures including controlled burns, clearing of dead wood, cutting tall grass and brush, grazing to thin vegetation in particular areas and similar measures.

- Brush clearance areas around structures and fuel breaks in strategic locations.

- Use of fire resistant roof and exterior wall materials.

- Adequate access roadways for emergency vehicles and exit roadways for residents.

- Water storage and distribution systems adequate for fire protection purposes.

- Development of exposure protection systems, incorporating technologies such as class A foam.[4]

**Disaster Response** – This incident was clearly beyond the capabilities of normal fire suppression forces, and a realistic approach to similar situations must be based on hazard mitigation and risk reduction. It is not feasible to provide a fire suppression organization to master the situation that occurred in Oakland. The fire departments in Oakland, Berkeley, and all of the surrounding areas provided a valuable lesson to the fire service, in demonstrating their courage, skills, and dedication. But they

---

[4] Residential automatic sprinkler systems alone would not provide protection against this type of fire exposure. Compressed-Air Foam Systems (CAFS) may prove to be the most effective exposure protection system for this type of severe exposure.

also demonstrated the need for risk assessment and planning for disasters that completely over-whelm regular emergency response systems. There are many things that can be said and done to react to these lessons:

- There are recognizable fire risk situations (not necessarily limited to wildland-urban interface environments) that are clearly beyond the capabilities of fire suppression forces. Hazard reduction strategies should be the primary approach taken when these situations are recognized.

- An urban conflagration resulting from a wildland interface fire is a situation that has not received sufficient planning attention.

- It is unrealistic to expect normal emergency response systems that are based on routine demands to smoothly manage a disaster situation; the test is how quickly and effectively a disaster response system can be implemented.

- It is impossible to manage a large scale disaster with insufficient command personnel and inadequate command and control systems. Fire departments should always anticipate "worst case" scenarios and develop plans and procedures to address those situations.

- Communications systems that are adequate for normal times and situations may be easily overwhelmed in a disaster situation.

- In the aftermath of a disaster, every detail of response operations is subject to review in min-ute detail. The inadequacies are grossly magnified.

- It is extremely difficult to evacuate a heavily populated interface zone, particularly when the homes are enveloped in rapid burning and easily ignitable fuels. When wind, terrain, narrow roads, steep grades, and other factors combine to accelerate fire spread and restrict passage, the risk to residents can be extreme. Once a fire starts spreading through the area, it may be too late to evacuate.

- Firefighters will subject themselves to extreme levels of risk and personal exertion in the efface of a disastrous fire. It may be necessary to order companies to evacuate for their own safety, or to rest when the situation is still out of control.

- When normal response resources are forced to retreat, it is extremely difficult to regroup, reorganize, and return to effective action. Command officers must recognize when condi-tions are failing and decommit early enough from futile operations in order to have time to regroup and reorganize effectively.

- Existing expectations for "jump potential" may need to be reevaluated. The distance that this fire jumped, from Hiller Highlands over the freeway interchange and Lake Temescal, was much faster than had been anticipated in planning for a fire in the hills. The conditions for a long "jump" were perfect and the fire spread rapidly into an entirely new area.

- It is not clear that even an early series of aerial attacks could have controlled this fire; however, the need for rapid notification and response of aircraft can be seen in the analysis. The arrival of helicopters and fixed wing aircraft to attack this fire was delayed by circumstances, com-munication problems, and confusion.

- It may have been feasible to protect some of the structures with exterior sprinkler systems, if adequate water flows and pressures had been available and the more severe exposures to wildland fuels had been reduced. The intensity of the exterior fire exposure was so severe that interior automatic sprinkler systems had no value in protecting the structures from ignition.

- Compressed Air-Foam Systems (CAFS) appear to be very useful in applying a thick foamy covering to protect severely exposed structures. There have been experiments with class A foam agents, applied through automatic sprinklers or deluge systems, or by handlines, as exposure protection systems. Research is being conducted on mobile (truck mounted) systems and fixed or portable CAFS systems for homeowners.

# APPENDIX A

# Reference Publications

**The East Bay Hills Fire**
A Multi-Agency Review of the October 1991 Fire in the Oakland/Berkeley Hills
East Bay Hills Fire Operations Review Group
State of California
Governor's Office of Emergency Services

**Hazard Mitigation Report for the East Bay Fire in the Oakland/Berkeley Hills**
(In response to the October 22, 1991 Federal Disaster Declaration Covering Alameda County, California)
FEMA-919-DR-CA

**Response of the San Francisco Fire Department to the Oakland Conflagration of October 20 and 21, 1991**
Edited by David Fowler
San Francisco Fire Department

**Preliminary Study of the 1991 Oakland Hills Fire and Its Relevance to Wood-Frame, Multi-Family Building Construction (NISTIR 4724)**
Kenneth D. Steckler, David D. Evans, and Jack E. Snell
Institute of Standards and Technology
U.S. Department of Commerce

**The Great Oakland, Los Angeles, and San Diego Fires, September 22 to 29, 1970**
Naval Ordinance Laboratory, White Oak
Silver Spring, Maryland
NOLTR 71-229

**The Oakland-Berkeley Hills Fire – October 20, 1991;** and *Wood Shingles – 1959* and
**The Devil Wind and Wood Shingles:**
The Los Angeles Conflagration of 1961
(Bel Air Fire)
By Rexford Wilson
Fire Protection Association
Quincy, Massachusetts

The California Department of Forestry and Fire Protection is conducting a detailed study of fire spread characteristics at the East Bay Hills Fire. This study included gathering detailed information on every structure that was destroyed, damaged, or exposed and is intended to yield a body of knowledge on the mechanisms that determine susceptibility to ignition in an interface.

# APPENDIX B

# East Bay Hills Fire Chronology on October 20th

Selected messages from the enormous amount of radio and telephone traffic that occurred on the first day of the fire are presented in this chronology to illustrate not only the sequence of events, but also the extreme complexity of this fire from the point of view of incident management.

0820    Off-duty battalion chief advises BC2 of strong wind condition developing in Oakland Hills area.

0830    BC2 sends E24 and E19 to check scene of previous day's fire.

0850    E19 and E24 on-scene at 7185 Marlborough Terrace locate spots inside fire line from previous day's fire. Hot spots noted in duff under pine trees on north flank and near Gwin Tank. EBRPD requested to respond to pick up their hose. EBRPD contacted by Oakland Fire Communications; unable to response until day personnel report for duty.

0908    BC2 orders Patrol 28 to be activated to increase fast response capability in Hills area.

0913    BC2 at scene at top of hill, at Marlborough Terrace near Grizzly Peak. Requests EBRPD to respond to assist with flare-ups inside burned area.

0916    BC2 requests additional engine company. E16 dispatched.

0925    EBRPD advised by Oakland Fire communications to respond and pick up hose or risk losing it to rekindled fire.

0926    BC2 advises Oakland Fire Communications of extreme fire hazard conditions in Hills area due to wind and fuel conditions.

0929    BC2 assumes command.

0932    EBRPD 5675 on-scene, reports that situation "seems to be OK." Advises additional units to respond Code 2 (non-emergency). EBRPD personnel assigned to work hot spots on lower north flank from Buckingham Boulevard.

0945    BC2 reports situation under control. Delegates command to E24.

        EBRPD chief advises to increase staffing at EBRPD fire stations due to unusual weather conditions.

0947    BC2 orders BC3 and BC4 to each assign one engine company to patrol Hills due to severe weather danger. E4 and E27 assigned.

0953    EBRPD Unit 5632 at scene, at Marlborough and Grizzly Peak.

0959    BC2 notes, "We have the most critical fire conditions in five years."

1004    EBRPD Unit 5675 at scene, along with two additional EBRPD firefighters.

1005    Companies at scene continuing to work hot spots. Hoselines set up at perimeter as precaution.

1015    E24 radios Oakland Fire Communications to advise EBRPD to have their units work hot spots on each flank. (EBRPD relays information to Unit 5675.) Oakland Fire Communications advises E24 to contact EBRPD units on scene on "White" (mutual aid channel).

1026    EBRPD advises Unit 5675 to contact Oakland units on "White" channel.

1029    E24 advises that E19 and E16 will be released, E24 and EBRPD will stay at scene with precautionary lines in-place.

1030    EBRPD personnel working spot fire with hand tools on west flank at bottom of burn area.

1035    Oakland Fire Communications advises E24 of telephone call from a resident at 7290 Marlborough Terrace reporting a hot spot on the hill. E24 reports this is the area they are working and EBRPD has advised that they can handle the situation.

        EBRPD 5675 advises 5632 to cover flare-up (open flame) on left flank. Engine 5632 positioned in driveway at 7151 Buckingham, extending line uphill to attack.

1037    EBRPD 5675 asks units if they can handle the flare-ups. Unit advises "10-4."

1040    Grass fire reported at 13685 Campus Drive, in Hills area approximately five miles south of this incident. E21, E25, and E27 dispatched. (Small grass fire was located and handled by E27.)

1041    E24 reports E19 will be in command. E24 will be leaving scene.

1050    EBRPD 5675 requests another EBRPD engine to respond Code 3. Reports "we're getting a lot rekindles up here." The EBRPD helicopter, Eagle 5, also requested.

1050-
1057*   E19 reports new flare-up with "pretty good smoke showing." BC2 directs E24 to return to assist E19.

        E19 reports smoke is coming from previously unburned area.

        Numerous spot fires and flare-ups on the slope with personnel moving quickly to keep them suppressed. E19 advises BC2 of the situation and reports it is still "under control."

        Radio traffic indicates a concern for the safety of personnel working in the area of flare-ups and difficulty maintaining radio communications between Oakland and EBRPD personnel.

1058    E19 requests full box alarm (first alarm) to respond to Gwin Tank. Reports "fire showing on the lower east flank." E24, E28, and BC2 dispatched. E4 advises he is in the area and also responding.

1059    E19 requests CDF response (acknowledged by Oakland Fire Communications).

1102*   E19 requests second alarm. Units to respond to 7140 Marlborough Terrace. Also requests Oakland Police Department for traffic and crowd control.

1104    BC2 en route, requests third alarm.

---

*Indicates an unverified time.

Also requests CDF on mutual aid; advises Oakland Fire Communications to inform CDF this is another fire in the "Threat Zone."

1105    Second alarm: E6, E10, T1, T15, BC4 dispatched.

1105-
1114    Oakland Fire Communications responds to medical incidents. Initial call made from Oakland Fire Communications to CDF. E19 reports fire crowding in trees below Marlborough. BC2 directs command van to set up at Tunnel entry. BC4 reports spot fire at Norfolk and Marlborough.

First structure becomes involved on Buckingham.

1115    BC2 requests fourth alarm. Assigns BC4 as Division A (DIV A) to supervise operations from Buckingham Boulevard.

1115-
1117*   E24 reports they think they have the fire cut off from spreading to the west from 7140 Buckingham.

E19 reports fire spreading along the north side of the canyon, behind houses on Buckingham.

E19 reports fire spreading on two fronts.

Off-duty assistant chief on-scene, advised to meet ComVan and establish command post at Tunnel entrance.

1118    Fourth alarm dispatched.

1119    CDF dispatches Helitack Copter 106 to Oakland.

CB2 requests confirmation that CDF is responding (no reply).

ConFire calls Oakland Fire Communications to ask if they have a fire; reply, "Same as yesterday."

1119-
1125*   Division A (BC4) reports fire is spreading rapidly toward Marlborough Terrace. Division A will be moving up to Marlborough from Buckingham.

Incident commander reports he cannot see fire spread from bottom of hill – just smoke.

BC2 assigns E6 and T15 to protect homes on Grizzly Peak Terrace.

BC2 advises need for additional command officer on top. E19 reports fire spread is lateral, not uphill. Fire spreading rapidly toward Norfolk Drive.

Division A reports structure fire and fire "going to jump Buckingham any minute."

At 1120, off-duty assistant chief assumes command of incident and requests fifth and sixth alarms.

BC2 request sixth alarm companies to respond to Bay Forest Drive and Tunnel Road. Also requests five engines on mutual aid to respond to Grizzly Peak Boulevard and Fish Ranch Road. "Ask Orinda if they can send us mutual aid to Grizzly Peak."

*Indicates an unverified time.

Command officers attempted to use Oakland Channel 1 but encountered E17 and T8 on McArthur Boulevard, who were using Oakland 1 for a working structure fire.

1123    ConFire dispatches first alarm to grass fire in Franklin Canyon (Martinez, approximately 20 miles northeast of Oakland fire).

1126    CDF Copter 106 en route to Oakland.

1129    CDF dispatches two engines, one dozer, and a battalion chief on mutual aid to Oakland, Grizzly Peak area.

1130    ConFire requests response from CDF for Franklin Canyon fire.

1132    Berkeley monitoring Oakland Fire Department radio traffic. Chief 6 (duty chief) responds to check Berkeley area. No immediate risk noted.

1133    Division A reports "Probably can't hold – it's coming over – we are abandoning task."

BC2 advises Division A to be extremely careful. "Don't get anybody killed!"

Incident commander requests five strike teams from Alameda County to stage at Hill and Tunnel Roads. Reports fire totally out of control and moving on several fronts involving more than 100 acres of trees, brush, and houses.

Pacific Gas and Electric requested to respond for live power lines down.

Helicopter Eagle 5 (EBRPD) attempting to contact Oakland Fire Communications on White 1.

California Highway Patrol transfer call to ConFire reporting fire on Orinda side, above Caldecott Tunnel. California Highway Patrol also advises ConFire to anticipate mutual aid request from Oakland.

1134    Command van redirected from top of hill to off-ramp at Tunnel entrance ("same place as yesterday").

Oakland Police requested to begin evacuating Parkwood Apartments. Fire is coming toward the complex.

BC2 advises Oakland Fire Communications to request Contra Costa County to respond, "fire is spreading to their area."

Division A reports "We're evacuating Buckingham – fire went over both sides of us."

E25 reports they are at 7160 Marlborough, protecting two houses.

Incident commander requests major response from Oakland Police Department. Need to evacuate 7100 block of Marlborough.

ConFire actives Plan 1.

ConFire dispatches first alarm brush assignment to Fish Ranch Road above Caldecott Tunnel entrance and advises Oakland Fire Communications they are responding.

CDF dispatches three engine companies and a battalion chief to Franklin Canyon on request from ConFire.

1136    ConFire dispatches second alarm to Franklin Canyon fire.

Alameda County advises ConFire of request from Oakland Fire Communications for one strike team.

1137 E1, T1, and T3 assigned to try to protect Parkwood Apartments. E6 reports fire is moving south from Grizzly Peak Terrace.

1138 ConFire advises Oakland Fire Communications they are responding.

1140 Eagle 5 attempting to contact incident commander on White 1.

1141 Incident commander attempting to find additional engine companies to assign to Parkwood. Companies attempting to evacuate complex. E16 assigned to assist.

1143 Orinda battalion chief (BC45) requests second alarm from ConFire to Fish Ranch Road above tunnel portals. ConFire dispatches.

ConFire contacts Oakland Fire Communications for instructions. Oakland Fire Communications requests two strike teams to respond to east tunnel entrance, then contact Tunnel Command on White 1. (Changed to west exit one minute later.)

Radio interference for several minutes on White 1 due to burning telephone line to Caldecott transmitter tower.

1144 Oakland's Division A reports "Fire at both sides – we're going to have to wait it out." (Last recorded communication from Chief Riley.)

Orinda BC45 advises ConFire that Oakland is requesting all available units.

ConFire calls San Ramon Valley dispatch and requests a strike team (ST2031A) to Caldecott Tunnel.

1145 E19 reports personnel trapped behind Gwin Tank. "Need help to get out, but not in imminent danger."

Alameda County logs first request for five strike teams from Oakland Fire Communications. Dispatches four task forces from Alameda County and requests one from ConFire. ConFire advises they have strike teams already en route.

1148 CDF Battalion Chief 1616 responding to Oakland advises Morgan Hill he can see both fires (Oakland and Franklin Canyon) and both are major.

1149 Orinda units trying to make a stand at top of hill above fire with Oakland companies; BC45 en route to Oakland command post – assigns captain as Division A.

1150 Training 2 advises of two structures involved at 7120 Norfolk. Oakland Fire Communications advises incident commander that seven strike teams are en route to the command van.

Hill Area Disaster Plan activated by incident commander. Agencies to prepare for mass evacuations.

1150-
1158 Incident commander requests estimated time of arrival for air support – need is critical. Oakland Fire Communications advises it has been ordered.

Patrol 20 advises they are at 7120 Marlborough with two houses involved. Fire has jumped street. They are protecting selves.

E1 reports rear buildings are becoming involved at Parkwood. Committing to interior rescue efforts.

Division A make several attempts to contact incident commander by radio – unsuccessful.

E8 reporting emergency conditions on Norfolk; needs assistance.

1153    CDF Copter 106 on scene, switching to White 1.

1155    Oakland requests mutual aid from Alameda City Fire Department.

1156    CDF dispatches Air Tankers 92 and 94 and spotter aircraft AA460 to Franklin Canyon fire. Planes respond from Salinas (90 miles) and Fresno (150 miles).

1157    ConFire dispatches third alarm to Franklin Canyon fire.

1159    Incident commander requests estimated time of arrival for air strikes; "fire is across tunnel."

CDF Battalion Chief 1616 requests aircraft status. Advised Copter 106 is only air unit on scene. Two air tankers en route to Franklin Canyon fire; all others committed to Sonoma County fire. Telephone call to ConFire from Orinda Fire Station 45, relaying message from Battalion Chief 45, to advise he needs air strike "right now – homes are going like crazy!"

1200    ConFire contacts CDF with air strike request. Advised no aircraft available. BC45 advised at 1204.

ConFire contacts Alameda County for situation update. Advises that ConFire will request an OES Strike Team.

1200*   Fire reported rapidly spreading through Parkwood complex, rear buildings fully involved. Companies attempting holding action while evacuation continues.

Command post relocated to Highway 24 and 13 interchange.

1201    E8 asking for "anyone to pump their line." Report Buckingham blocked by parked cars and downed power lines – "need help!"

Incident commander advises "waiting for strike teams to arrive – everyone is committed."

1201    ConFire calls OES Region 2. Phone is forwarded to CDF Santa Rosa. Santa Rosa will advise OES 2 to pick up calls.

1202    E2 reports fire in trees at Temescal Recreation Area; no structures involved at this time.

E8 reports "EMERGENCY – power lines down on hydrant!" Needs Pacific Gas and Electric.

ConFire calls Crane Helicopter (private contractor for water drops) to request their response.

1203    East Bay Municipal Utilities District (EBMUD) calls ConFire for information on fire situation.

CDF Battalion Chief 1616 on scene, Fish Ranch Road.

1204    ConFire contacts OES Region 2. Advises OES of situation with two major fires, requests two strike teams to stage at Orinda and Concord fire stations.

---

*Indicates an unverified time.

1206    Crockett Fire District first alarm response assisting at Franklin Canyon fire.

1207    Berkeley E3 dispatched to 32 Vicente Drive on citizen report of fire coming over the hill above that address.

    BC45 contacts ConFire to see if they are in contact with Copter 106. Reply is negative. Also ConFire is unable to contact Oakland Fire Communications.

1209    CDF Engine 1161 on scene at Tunnel Fire (first CDF engine to arrive).

1210    Oakland Fire Communications advises incident commander, "Tanker and helicopters en route." Incident commander reports "Fire moving fast on multiple fronts – losing structures."

    Moraga Task Force at scene, assigned to assist units on Grizzly Peak Terrace.

1210    ConFire contacts Richmond Fire Department for mutual aid, two engines to cover Orinda.

1211    Telephone discussion between ConFire assistant chief and CDF division chief at Morgan Hill, updating situation. ConFire suggests diverting aircraft from Franklin Canyon to Oakland due to structures and report from BC45; Franklin Canyon is the lesser threat.

    CDF advises Contra Costa that Oakland Fire Communications has only requested mutual aid response (two engines, one dozer, and one battalion chief). Call continues until 1222.

1212    BC44 (off-duty battalion chief) reports houses involved on Contra Costa Lane west of Lake Temescal. Establishing command post at Contra Costa and Buena Vista with E2.

    Strike Team 2031A en route from San Ramon to Tunnel.

1213    Berkeley E3 checks area on Vicente Drive, continues to Grandview and Westview before seeing fire on the ridge at approximately Norfolk/Buckingham area.

    E3 requests Berkeley first alarm. E2, E5, T5, Chief 6, and Paramedic 113 dispatched.

1215    Engine 8 reports no water coming from hydrant on Buckingham. CDF Helicopter 106 working with BC44.

    E3 and E13, assigned to assist BC44, stage at Broadway and Golden Gate.

    CDF Engines 1661 and 1674 on scene at Grizzly Peak and Marlborough Terrace; assessing situation and trying to determine if fire is spreading to State responsibility area.

1217    CDF asks E1676 at Franklin Canyon fire if air tankers can be diverted.

1218    BC44 suggests requesting mutual aid from San Francisco. Incident commander orders Oakland Fire Communications to request ten engine companies from SFFD.

1220    Structure fire reported on Country Club Drive.

1220-
1230    CDF E1616, E1661, and E1674 working with Oakland companies trying to save structures on Marlborough Terrace. Main body of fire had already gone past this area, headed southwest, pushed by 10-30 miles per hour wind. CDF and Oakland units unable to establish radio contact with Oakland command post.

1222    Division A attempting radio contact with BC44 – unsuccessful.

1223    Oakland Fire Communications advises incident commander, six tankers and six helicopters en route. Report of invalid trapped in home at 1616 Northhill.

1223    CDF Morgan Hill asks Franklin incident commander if air tankers can be diverted.

1225    Berkeley E2 assumes command at Vicente Road. Fire coming over the hill above with flying brands starting several fires ahead of the flame front. Companies committed to prevent involvement of structures.

        Franklin incident commander approves diverting air tankers to Oakland.

1226    ST3031A calls ConFire on cellular phone and reports unable to contact Tunnel incident commander by radio. Advised to go up Fish Ranch Road to find them.

        CDF battalion chief advised Morgan Hill that he is trying to organize CDF resources on north side of fire; unknown number of acres and structures involved, but fire is moving rapidly in all parts with 15 miles per hour winds. Having difficulty organizing situation.

1229    SFFD receives first request for assistance from Oakland.

        SFFD Strike Team 1 (E1, E3, E8, E29, E36, and BC3) assembles at west end of Bay Bridge.

1230    CDF units still unable to contact incident commander. Determine that CDF units will take independent action to limit fire spread to north toward Claremont Avenue.

1231    Berkeley incident commander request second alarm. E1, E3, T2 dispatched.

        CDF attempts to contact aircraft en route to Franklin fire.

1233    Berkeley units begin evacuating Vicente Road.

1236    CDF contacts Air Tankers 92 and 94 and Air Attack 460 and diverts them to Oakland.

1237    Radio traffic between Orinda E44 and Division A, "about to be surrounded."

1239    Additional telephone contact between ConFire assistant chief and CDF division chief regarding locations and authorizations for air strikes. Oakland Fire Communications has requested only one helicopter. CDF will dispatch a third air tanker. Air Tanker 77 dispatched from Ukiah (estimated time of arrival 1316 hours).

1240    SFFD Strike Team 2 (E6, E7, E13, E17, E25, and BC9) assembles at west end of Bay Bridge.

1240*   Berkeley E5 and TS assigned to Chabot Road area to check for fire extension. Fires discovered starting in trees and spreading to structures.

1242    Spotter Aircraft A4460 overhead requesting CDF BC1616 to assign air-to-ground frequency.

1244    E44 advises ConFire of no water pressure at Grizzly Peak and Fish Ranch Road. ConFire to advise EBMUD.

1245    ConFire officer at Franklin Canyon requests additional strike team. Fire has jumped fire break. Relayed to CDF.

1245*   Berkeley companies driven out of Vicente Road. Master streams set up to make stand along Tunnel Road, Bridge Road, and Alvarado Road ahead of fire. Area designated as Berkeley Division A. Chabot Road is designated as Division B.

---

*Indicates an unverified time.

1249    Orinda E44 attempting to contact Division A or Division B.

1250    AA460 advises position and altitude for air operations over Tunnel fire. Requests air space clearance from Oakland Airport Control Tower.

1253    ConFire advises E44 that EBMUD is aware of no water situation; they are having power problems. EBMUD wants to know if roads are clear to access pumping station.

        E42 reports to ConFire, "Oakland chiefs say they are going to pull out of area."

1259    Air Tankers 92 and 94 begin first slurry drops. (At 1303 en route to Santa Rosa Airport to reload.)

1300    CDF units proceed to Claremont Canyon and take Alvarado Road from east end. Encounter structure fires starting up around Amito Road.

        SFFD ST2 assigned to protect Claremont Hotel. Deployed in defensive positions under Oakland Division C (Captain Parker).

1308    Second Contra Costa Strike Team (ST2036A) en route from Walnut Creek.

1309    Alameda County Mutual Aid --

        Oakland requests:
        Three strike teams to Grizzly Peak (two Type 1, one Type 3)
        One Strike Team to Golden Gate and Acacia (Type 3)
        Three command officers requested to command post.

        Berkeley requests:
        Two strike teams (Type 1) to stage at Berkeley High School.

1313    BC45 advises ConFire that Oakland command post is now at Lake Temescal.

1315    CDF Copter 96 en route to Tunnel fire.

        ConFire receiving multiple reports of major power lines exploding and starting fires on Contra Costa side of hills, one in Lost Valley, Moraga and one near Brookwood Apartments, Orinda. Both are threatening structures.

1318    ConFire dispatches Orinda E45 and four Richmond units to Los Valley for grass fire (Dolores fire).

1319    AA460 advises CDF of two new fires started by power lines shorting. One fire is 1 mile south, second is 2 miles south of main fire (Dolores and Sunrise fires).

1320    Citizen calls ConFire to advise of Lost Valley fire; reports two helicopters are approaching.

        CDF 1616 calls Morgan Hill on cellular phone to report on situation.

1321    ConFire diverts ST2036A to Orinda.

1327    ST2031A leader advises ConFire – unable to contact any Oakland units; ST is protecting structures, need to know if Oakland units are pulling out.

1329    Air Tankers 94 and 92 land at Santa Rosa to reload.

---

*Indicates an unverified time.

1330    CFD 1616 reports multiple structures involved and brands starting new fires in Alvarado Road area. Some residents fleeing; others on rooftops with hoses.

1330*   Alameda County Task Force 13 en route to Golden Gate and Acacia, stopped at Broadway Terrace and Proctor to rescue resident from endangered residence. Fire crossing Broadway Terrace at this location.

1330*   Piedmont E1 and E2 respond to area of Florence and Modoc to engage fire heading toward city limits on direction of Piedmont Chief. Units committed in this area until approximately 0100 hours.

1334    CDF dispatches five engines, one dozer, and BC1611 to brush fire in Bollinger Canyon.

1337    ConFire dispatches Moraga E42 to grass fire on Brookwood Road; under control at 1347.

1342    Alameda Fire Department ambulance requested to EMS staging area.

1346    ConFire dispatches three OES engines (part of ST2803A) from Orinda to Dolores fire.

        Franklin fire incident commander advises "shaky containment."

1350*   TF13 unable to reach assigned location. Self-assigned to attempt to stop progress along Proctor, Florence, Agnes, and Modoc. Battle continues in this area for six hours.

1351    ConFire dispatches E42 to Sunrise fire. (Fire eventually requires nine units to control.)

1352    CDF diverts units responding to Bollinger Canyon to stage as strike team (ST9160C) at Sunol.

1355    CDF directs BC1612 from Franklin fire to Tunnel fire; en route at 1410.

1359    Oakland requests:

        Seven Type 1 Strike Teams, and
        Six Type 3 Strike Teams to stage at Raimondi Park
        Six additional air tankers
        Six additional helitack units.

1400    SFFD command staff alerted. Smoke and burning embers reported reaching San Francisco.

1400*   Berkeley assistant chief assigned as liaison at Oakland command post.

1410    SFFD OES E217 placed in-service to respond to staging area in Contra Costa County. (Later responds to Oakland.)

1414    Alameda E3 requested to cover Oakland Station 17. Later dispatched to Hills fire from FS17.

1415    CDF BC1616 directed by Morgan Hill to report to Oakland command post at Hiller and Tunner Roads.

1416    ConFire calls Pinole, Crockett, Richmond to assemble additional strike team to stage at Station 15, Lafayette.

1430    Berkeley requests two additional strike teams (Type 1) to stage at Berkeley High School.

        CDF reassigns four additional air tankers from Geysers fire north of Santa Rosa to Oakland.

---

*Indicates an unverified time.

1449   ConFire advised Franklin incident commander will begin releasing Contra Costa units.

1452   BC1612 on-scene at CDF staging area – Fish Ranch and Grizzly Peak. Directs ST9160C to his location at 1457.

1500   CDF BC1616 locates Oakland command post on Highway 24. Establishment of unified command begins.

1506   CDF advises AA440 of aircraft assigned to incident: seven air tankers and four copters.

1513   ST2037A assembled at Contra Costa Station 15, Lafayette (dispatched to Oakland at 2028).

1520   ConFire assembles ST2040A and dispatches to Dolores fire.

1522   SFFD activates HT8 and HT15 with crews of T8 and T15. Units respond to Oakland with high volume relay/supply hose.

1527   AA440 assigns AA140 to control air operations over Dolores and Sunrise fires.

1532   ConFire dispatches three additional engines to Tunnel fire.

1540   Oakland requests two caches of 20 portable radios each from Alameda County.

1545   CDF BC1612 at command post, developing overall plan for incident management with Oakland command staff.

1546   CDF BC1614 and 1602 en route to establish communications at Raimondi Park staging area.

1548   ConFire dispatches ST (PW5, 6, 19, 51, and PT38) to Dolores fire.

1549   SFFD dispatches three water tenders and a command van from the Department of Public Works to Oakland.

1552   CDF assigns additional (eighth) air tanker.

1555   CDF air units having difficulty making contact with incident commander on ground at Dolores fire.

1556   Contra Costa units working in area of Claremont and Alvarado, shuttling water, trying to protect structures.

1558   ConFire dispatches ST (51A, 54A, 73, and 78) to Dolores fire.

1600   Incident commander and CDF BC1616 determine air attack priorities as both sides of Highway 13 south of Highway 24. BC1616 radios instructions to aircraft.

1604   CDF assigns additional (ninth) air tanker.

1614   ConFire dispatches ST2041C to Dolores fire.

1615   ST9160C (CDF) deployed along Claremont Avenue.

1617   Franklin fire incident commander begins releasing CDF units.

---

*Indicates an unverified time.

1700    CDF BC1616 surveying Branch 2; Broadway Terrace from Skyline to Highway 13. Determines Broadway Terrace is viable cut-off point. Locate Oakland and EBRPD crews working in this area.

Branch 1 is Tunnel Road to Grizzly Peak along Claremont.

Branch 2 is south and east perimeter of fire to Highway 13.

Branch 3 is Rockridge District to Claremont Hotel.

1715    SFFD dispatches 60 additional personnel by bus to assist at the Claremont Hotel.

1727    Sunrise fire contained at 200 acres, one structure.

1746    Oakland requests eleven additional strike teams: Five Type 3 and Six Type 1.

1815    SFFD requested to provide ten additional strike teams. Unable to meet request; however, 25 additional personnel sent to Oakland by bus.

1930    OES Region 2 requests ST2043A from ConFire to Oakland.

2002    ConFire releases San Mateo Strike Team to Oakland.

2028    OES Region 2 requests ST2037A from ConFire to Oakland.

2030    ConFire ST2031A released from Oakland.

2045    ConFire ST2041C released from Dolores fire.

2050    Oakland requests thirty additional strike teams: 20 Type 1 and 10 Type 2.

2100    ConFire ST2038A and 2040A (PW 69, 15, E15A, 36A) released from Dolores fire.

2136    ConFire first and second alarm companies released from Oakland.

2145    ConFire ST2036A released from Oakland.

2200*   OES Overhead Team arrives at command post.

2220    Oakland requests ten additional Helitack units, planned need for dawn.

2241    OES Region 2 requests ST2804A from ConFire to Oakland.

2245    OES Region 2 requests ST2042A from ConFire to Oakland.

2300*   Staging area relocated to Alameda Naval Air Station.

0304    Berkeley requests two additional strike teams for 0800 hours.

---

*Indicates an unverified time.

# APPENDIX C

## Contra Costa County Summary

This appendix is provided as a supplement to demonstrate the complicated situations that were experienced throughout the area on the day of the fire.

Contra Costa County Fire Protection District (ConFire) became aware of the East Bay Hills fire from callers reporting smoke coming over the hills. ConFire called the Oakland Fire Department Communications Center at 1119 hours. Oakland confirmed that they had a major fire in the same location as the previous day.

A major brush fire was reported in the Franklin Canyon area of Contra Costa County, approximately six miles northeast of the Oakland fire, at 1123 hours. ConFire immediately dispatched a first alarm and requested a full response from CDF at 1130 for the Franklin Canyon fire. CDF dispatched three engine companies and a battalion chief at 1135 hours. One minute later a second alarm was requested from ConFire for the Franklin Canyon fire.

ConFire dispatched a first alarm assignment to Fish Ranch Road at 1135, when a citizen calling from a cellular telephone reported a fire on the hillside above the east portal of the Caldecott Tunnel. Units from the Orinda and Moraga Fire Departments were included in this assignment, along with Contra Costa County Fire Department companies from the Lafayette station, under Orinda Battalion Chief BC45.

At 1136, Alameda County called ConFire to advise that Oakland was requesting a strike team from Contra Costa County. ConFire immediately called Oakland to inform them that a full first alarm response was already en route. A second alarm was requested by BC45 at 1143 and additional companies were dispatched from Orinda, Lafayette, and Walnut Creek.

The ConFire had some difficulty reaching the Oakland Communications Center to clarify the request for a strike team. Contact was made at 1144, and Oakland requested two strike teams, one wildland and one structural, to respond to the Caldecott Tunnel and contact Oakland on White Channel. At the same time, BC45 was reporting to ConFire that Oakland was requesting "all available units." ConFire immediately called up Strike Team 2031A consisting of San Ramon Valley and Dougherty companies. This strike team was en route at 1212 and in the fire area at 1226 hours.

Unable to reach the incident commander by radio, the Orinda Battalion Chief assigned a captain as "Division A" and directed the incoming units respond to Fish Ranch and Grizzly Peak, while he went to make direct contact at the Oakland command post.

At 1156 air support for the Franklin Canyon fire was dispatched by CDF, consisting of a spotter aircraft and two slurry bombers. A third alarm was requested at 1157, and additional Contra Costa County units were dispatched. At 1206 the Crockett Fire Department called to advise ConFire that they had a first alarm assignment staged near the Franklin fire. These units were also requested to assist and were designated as Division B on the Franklin fire.

At 1159 ConFire received a telephone call from Orinda Fire Station 45, relaying a message from BC45, that air strikes were urgently needed on the Oakland fire; "Right now homes are going up like crazy." ConFire called CDF to relay the request and was told that no aircraft were available. This information was given to BC45 and 1204 hours. At 1202 hours, ConFire called Crane Helicopter, a private water drop contractor, and activated them for the Oakland fire.

At 1204 hours, ConFire called OES Region II requesting two strike teams to cover Contra Costa County. One was requested to respond to Station 45 in Orinda and one to Station 10 in Concord. At 1210, ConFire asked the Richmond Fire Department to send two engine companies, Code 3, to cover Station 45.

At 1215 a telephone conversation was initiated between the Contra Costa Assistant Chief and the CDF Division Chief at Morgan Hill. They discussed the situation at both fires and the urgent need for air support at the Oakland fire. Contra Costa gave an order to CDF requesting air support for the Hills fire and suggested diverting the aircraft from Franklin Canyon to Oakland, based on the threat to structures. (At the time this conversation was taking place, the first two CDF engines had just arrived at the Franklin fire and two CDF engines had been on the scene in Oakland for approximately five minutes.) The aircraft were diverted at 1236, after Morgan Hill verified the situation with the CDF-incident commander at Franklin Canyon.

At 1222, ConFire called up Strike Team 2036A, consisting of units from Riverview, Byron, East Diablo, and Oakley, to assemble in Riverview and respond to Concord Station 10. This strike team was assembled and en route to Oakland at 1255.

At 1233, OES Region II called ConFire to authorize activation of two OES engines from El Cerrito and San Ramon for Strike Team 2803A, to respond to Orinda. Additional OES engines from Hollister and Benicia were to meet them at Station 45. A San Mateo County strike team was also being sent by Region II to cover Contra Costa County.

At 1248 CDF determined that the Franklin Canyon fire was in a State responsibility area and assumed command of that incident. At that point CDF had three engines on the scene, with four additional engines, two hand crews, two dozers, and a battalion chief en route. Contra Costa County had a three-alarm assignment on the scene, assisted by a one-alarm assignment from Crockett.

## ADDITIONAL FIRES

At approximately 1315 hours, the high voltage power lines shorted out in the area of the Oakland fire, igniting two additional fires on the Contra Costa side of the hills, approximately one mile and two miles southeast of the Oakland fire. ConFire began receiving multiple calls on these fires, giving locations in Orinda and Moraga.

ConFire originally dispatched Orinda and Richmond units to a fire in the area of the Pacific Gas and Electric substation in Moraga, which was named as the Dolores fire. ConFire also diverted ST2036A from responding to Oakland to stand-by at Station 45. Four helicopters that were working the Oakland fire spotted the new outbreaks and responded to them.

At 1334 hours, CDF dispatched a full wildland response to Bollinger Canyon in Contra Costa County. This was probably a duplicate report on the Moraga fire. These companies were rerouted at 1352 to assemble at Sunol and then responded to the Oakland fire as ST9160C at 1452.

Several additional units from Contra Costa County were dispatched to the Dolores fire in Moraga, including three of the EOS engine companies from ST2803A and companies from ST2036A. Region II assigned additional EOS engine companies to reassemble ST2803A, which then responded to Oakland.

The second new fire was identified as the Sunrise fire in Orinda. At 1351 hours, companies from Orinda and Moraga responded to this incident. They were assisted by some of the Contra Costa County and Richmond companies that were standing by in Orinda. A total of nine companies were used on this fire which was reported to be contained at 1727 hours.

At 1346, the Franklin fire was reported to be contained and at 1449 hours the Contra Costa County units were ready to turn the scene over to CDF for overhaul. Several of the Contra Costa County units were reassembled into ST2040A and responded to the Dolores fire at 1548 hours. AT 1415, ConFire had called up a strike team (ST2038A) consisting of units from East Diablo, Pinole, Crockett, and Rodeo to assemble at Station 15 in Lafayette. This strike team was assigned to the Dolores fire at 1558 hours. An additional strike team, consisting primarily of Contra Costa County Fire Department companies (ST2041C) was activated and sent to the Dolores fire at 1614 hours. A total of approximately 49 ground units and four helicopters were used to control the Dolores fire by 2047 hours.

Strike Team 2037A, consisting of units from Riverview, Oakley, Bethel Island, and Richmond was called up at 1513 and was staged at Station 15. As units were released from the other fires, this strike team was dispatched to Oakland at 2028 hours. The San Mateo County strike team was released to respond to the Oakland fire at 2002 hours.

Contra Costa County supplied three additional strike teams to Oakland, on requests from OES Region II. Strike Team 2043A was called up at 2045 hours and consisted of units from East Diablo, Rodeo, Pinole, Oakley, and Crockett. A strike team of OES engines from San Ramon, Riverview, Orinda, and El Cerrito (ST2804A) was sent to Oakland at 2241 hours, followed by ST2042A, consisting of El Cerrito, Riverview, Crockett, Richmond, and Contra Costa County units. Several of the units included in these strike teams had seen action in the earlier incidents.

The first Contra Costa units were released from the Oakland fire before midnight on October 20, as fresh resources were brought in to relieve the most fatigued units. Some of the later responding strike teams were held in Oakland for several days.

# APPENDIX D

# Strike Teams

To illustrate the complexity of the situation, with multiple, simultaneous, major incidents in both Alameda and Contra Costa Counties, the following is a partial summary of resource deployments involving Contra Costa County for part of October 20, 1991. (This information was compiled from a number of different sources and may not be completely accurate. It is as accurate a log as could be obtained from the multiple information sources that were consulted.)

## Contra Costa County

| 1135 | First Alarm | BC45 | Orinda |
|---|---|---|---|
| | | E43 | Orinda |
| | | PW16 | Contra Costa (Lafayette) |
| | | PW42 | Moraga |
| | | Tal5 | Contra Costa (Lafayette) |
| | Released at 2136 | | |

| 1143 | Second Alarm | PW45 | Orinda |
|---|---|---|---|
| | | E44 | Orinda |
| | | PW17 | Lafayette ConFire |
| | | E4 | Walnut Creek ConFire |
| | | Tal | Walnut Creek ConFire |
| | Released at 2136 | | |

| 1212 | ST2031A | BC31 | San Ramon |
|---|---|---|---|
| | | E38 | San Ramon |
| | | E31 | San Ramon |
| | | E34 | San Ramon |
| | | E35 | San Ramon |
| | | E142 | Dougherty |
| | Called up by ConFire at 1144 | | |
| | En route to Oakland at 1212 | | |
| | Arrived area at 1226 | | |
| | Released at 2145 | | |

# Appendix D (continued)

| 1255 | ST2036A | BC81 | Riverview |
|---|---|---|---|
| | | E81 | Riverview |
| | | E97 | Byron |
| | | E51 | East Diablo |
| | | E52 | East Diablo |
| | | E94W | Oakley |

Ordered up by ConFire at 1222, Assemble at Station 10
Dispatch noted CAD at 1255
En route to Oakland at 1308
Held at Orinda at 1321
Released from Oakland at 2145

| 2028 | ST2037A | BC81A | Riverview |
|---|---|---|---|
| | | E82 | Riverview |
| | | E97 | Byron |
| | | E93 | Oakley |
| | | E93A | Oakley |
| | | E95A | Bethel Island |
| | | E62 | Richmond |

Assembled at Lafayette FS15 at 1513
Dispatched to Oakland at 2028
En route at 2034

| 1440 | ST2038A | AC5101 | East Diablo |
|---|---|---|---|
| | | E51A | East Diablo |
| | | E73 | Pinole |
| | | E78 | Crockett |
| | | E54A | East Diablo |
| | | E72 | El Cerrito |
| | | E76 | Rodeo |

Called up by ConFire to Lafayette FS15 at 1440
To Dolores fire at 1558
On scene at 1619
Released at 2100

| 1520 | ST2040A | 1170 | Contra Costa |
|---|---|---|---|
| | | E11 | (reassigned to ST2041C) |
| | | PW5 | Contra Costa |
| | | PW6 | Contra Costa |
| | | PW19 | Contra Costa |
| | | PT38 | Contra Costa |
| | | PW51 | East Diablo (sub for Ell) |

Sent to Dolores fire at 1548
Released at 2100

# Appendix D (continued)

| 1614 | ST2041C | 1113 | Contra Costa |
|---|---|---|---|
| | | E11 | Contra Costa |
| | | PW1 | Contra Costa |
| | | PW10 | Contra Costa |
| | | TA12 | Contra Costa |
| | | PW44 | Contra Costa |

To Dolores fire at 1614

| 2236 | ST2042A | BC71 | El Cerrito |
|---|---|---|---|
| | | E81A | Riverview |
| | | E78 | Crockett |
| | | E10C | Concord ConFire |
| | | E10A | Concord ConFire |
| | | E61 | Richmond |

E78 to FS15 at 1418
En route to Oakland at 2236
Released from Oakland at 1300 October 24th

| 2045 | ST2043A | AC5101 | East Diablo |
|---|---|---|---|
| | | E75A | Rodeo |
| | | E73 | Pinole |
| | | E94WT | Oakley |
| | | E79 | Crockett |
| | | E51 | East Diablo |
| | | E73 to FS15 at 1416 | |

Dispatched to Oakland at 1930
En route at 2045
Released at 1541 on October 23rd

| 1222 | ST2803A | OES199 | El Cerrito |
|---|---|---|---|
| | | OES233 | San Ramon |
| | | OES | Hollister |
| | | OES | Benicia |

Called up by OES at 1233 per ConFire request
Dispatched to fill-in at FS45 Orinda
Split-up between Dolores and Sunrise fires

*Reconstituted with OES engines from*

| | | | San Francisco |
|---|---|---|---|
| | | | San Mateo |
| | | | Pacifica |
| | | | Kenfield |
| | | | Ross Valley |

# Appendix D (continued)

| 2241 | ST2804A | BC3113 | San Ramon |
|------|---------|--------|-----------|
|      |         | OES159 | Lawrence Livermore  Laboratory |
|      |         | OES140R | Riverview |
|      |         | OES235 | San Ramon |
|      |         | OES237 | Orinda |
|      |         | OES199 | El Cerrito |

*Also responded to Oakland from Contra Costa County:*

| E71, PW71, PC7120 | El Cerrito |
|-------------------|------------|
| E93 | Oakley |
| E45 ComVan 45 | Orinda |
| E64 | Richmond |
| E32 | San Ramon Valley |

# APPENDIX E
## Photographs

The photographs that follow were taken one week after the fire by J. Gordon Routley except where otherwise noted.

# Appendix E (continued)

View looking toward the area of fire origin – on the grassy slope in the background. The left flank of the fire spread laterally behind the homes on Buckingham Boulevard and up toward the area in the foreground.

# Appendix E (continued)

View of the area of fire origin as seen from the Parkwood Apartments. The Saturday fire burned up the grassy slope on the hills toward Marlborough Terrace and the radio tower on Grizzly Peak. The fire on Sunday broke out in the lower part of the original burn area.

# Appendix E (continued)

One of the few homes on Marlborough Terrace that was not destroyed by the fire. Houses on both sides burned to the ground.

## Appendix E (continued)

All that remains of one hillside home is the steel supporting structure for the main floor.

# Appendix E (continued)

Buckled steel beams mark the location of a destroyed home on Buckingham Boulevard.

# Appendix E (continued)

One of the few surviving structures on Buckingham Boulevard, near the area of fire origin. This is the house where firefighters and residents took refuge, using all available water flow to protect the surface as the fire front passed over.

# Appendix E (continued)

Looking down on the ruins of the Parkwood Apartments from Buckingham Boulevard. Tunnel Road is visible in the lower left corner.

# Appendix E (continued)

Ruins of the Parkwood Apartments; previously four-story buildings.

# Appendix E (continued)

Ruins of the Parkwood Apartments; previously four-story buildings.

# Appendix E (continued)

Total destruction of Hiller Highlands complex.

# Appendix E (continued)

Total destruction of Hiller Highlands complex.

# Appendix E (continued)

Total destruction of Hiller Highlands complex.

# Appendix E (continued)

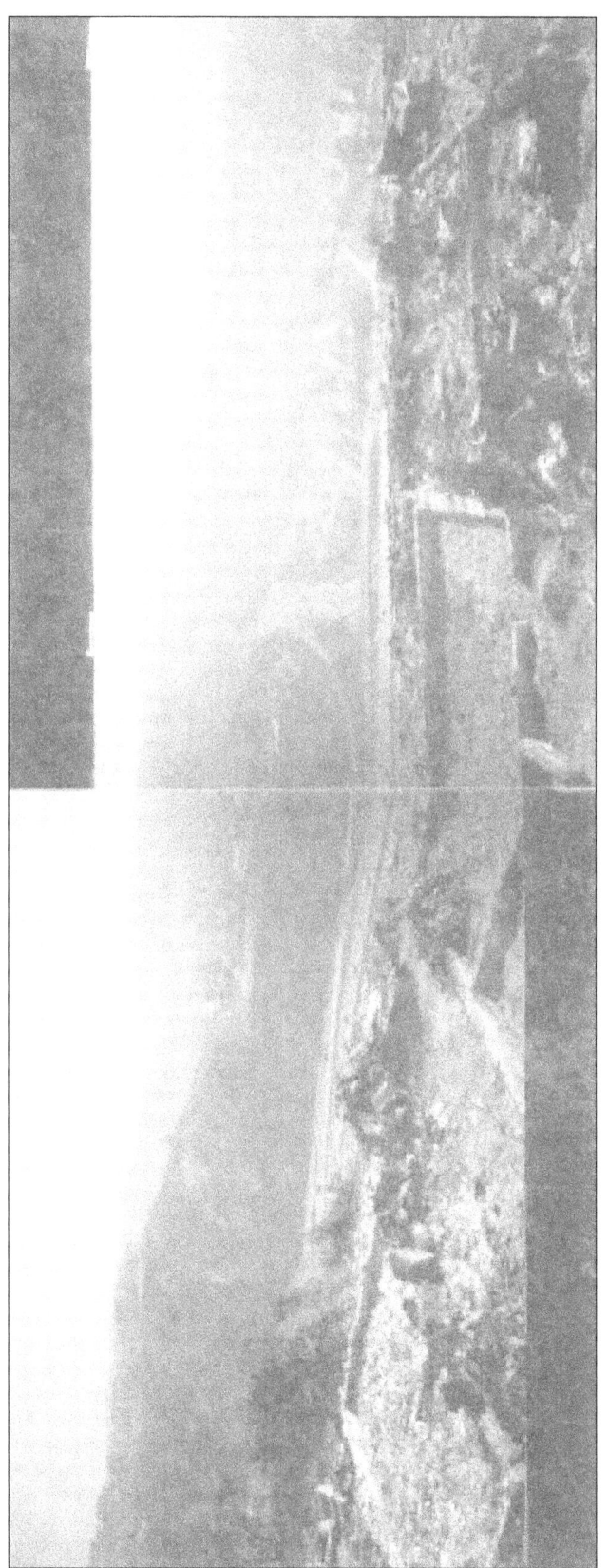

View from Hiller Highlands, looking down on interchange of Highway 24 and Highway 13. Entrance to Caldecott Tunnel is to the left. Lake Temescal and Rockridge District in the background. Fire jumped from Hiller Highlands to the trees beyond Lake Temescal, a distance of approximately 2,000 feet.

# Appendix E (continued)

Ruins of large two-story home in Rockridge district.

# Appendix E (continued)

Huge pine trees in the Rockridge district ignited and accelerated the fire spread in this area.

# Appendix E (continued)

Examples of development in the fire area – wood shingle roofs.

# Appendix E (continued)

Typical homes in Alvarado Road area.

# Appendix E (continued)

Home enveloped in natural vegetation.

# Appendix E (continued)

Homes enveloped in natural vegetation – fire area in background.

# Appendix E (continued)

Typical homes on western edge of fire area.

# Appendix E (continued)

Homes saved within fire area – note non-combustible roofs.

# Appendix E (continued)

Home with wood shingle roof destroyed. Neighboring home with tile roof saved.

# Appendix E (continued)

Home with wood roof burned off.

# Appendix E (continued)

Large home saved by firefighters after vegetation became involved.

# Appendix E (continued)

Large home saved by clear area around structure.

# Appendix E (continued)

West limits of fire area – groups of homes saved, others lost.

# Appendix E (continued)

The Claremont Hotel – Five-story wood frame structure in the path of the fire, considered a "conflagration breeder."

# Appendix E (continued)

One of approximately 2,000 burned automobiles in the fire area.